Air and Missile Defense Systems Engineering

Air and Missile Defense Systems Engineering

Warren J. Boord
John B. Hoffman

CRC Press
Taylor & Francis Group
Boca Raton London New York

CRC Press is an imprint of the
Taylor & Francis Group, an **informa** business

CRC Press
Taylor & Francis Group
6000 Broken Sound Parkway NW, Suite 300
Boca Raton, FL 33487-2742

© 2016 by Taylor & Francis Group, LLC
CRC Press is an imprint of Taylor & Francis Group, an Informa business

No claim to original U.S. Government works

Printed on acid-free paper
Version Date: 20160217

International Standard Book Number-13: 978-1-4398-0670-8 (Hardback)

This book contains information obtained from authentic and highly regarded sources. Reasonable efforts have been made to publish reliable data and information, but the author and publisher cannot assume responsibility for the validity of all materials or the consequences of their use. The authors and publishers have attempted to trace the copyright holders of all material reproduced in this publication and apologize to copyright holders if permission to publish in this form has not been obtained. If any copyright material has not been acknowledged please write and let us know so we may rectify in any future reprint.

Except as permitted under U.S. Copyright Law, no part of this book may be reprinted, reproduced, transmitted, or utilized in any form by any electronic, mechanical, or other means, now known or hereafter invented, including photocopying, microfilming, and recording, or in any information storage or retrieval system, without written permission from the publishers.

For permission to photocopy or use material electronically from this work, please access www.copyright.com (http://www.copyright.com/) or contact the Copyright Clearance Center, Inc. (CCC), 222 Rosewood Drive, Danvers, MA 01923, 978-750-8400. CCC is a not-for-profit organization that provides licenses and registration for a variety of users. For organizations that have been granted a photocopy license by the CCC, a separate system of payment has been arranged.

Trademark Notice: Product or corporate names may be trademarks or registered trademarks, and are used only for identification and explanation without intent to infringe.

Visit the Taylor & Francis Web site at
http://www.taylorandfrancis.com

and the CRC Press Web site at
http://www.crcpress.com

Contents

List of Figures ix
List of Tables xv
Preface xvii
Authors xix

1. Introduction and Background 1
   1.1   Introduction 1
   1.2   Why Is Missile Defense an Important Topic? 1
   References 2

2. Systems Engineering Fundamentals 3
   2.1   Pre–Phase A 7
   2.2   Phase A 8
   2.3   Phase B 9
   2.4   Phases C–F 11
   2.5   Concept-of-Operations Development 12
   2.6   Product Architecture 13
   2.7   Requirement-Driven Acquisitions 14
   2.8   Verification and Validation 17
   2A   Appendix: Systems Engineering Management Plan 17
       2A.1   Background 17
       2A.2   SEMP Outline 18
              2A.2.1   Introduction 18
       2A.3   Integrated Product Team Structure and
              Responsibilities 19
              2A.3.1   Technical Management 19
              2A.3.2   The Development of the Architectural
                            Design Solution 19
              2A.3.3   Product Development 20
       2A.4   The Milestone Review Process 20
       2A.5   Requirements Identification and Analysis Process 20
              2A.5.1   Requirements Identification 20
              2A.5.2   Requirements Management 21
       2A.6   The Verification and Validation Process 21
              2A.6.1   Verification 21
              2A.6.2   Validation 21
       2A.7   Configuration Control Management Process 22
       2A.8   The Risk Management Process 22
   References 22

3. Missile Defense Problem 23
3.1 Overview of the Missile Defense Problem 23
3.2 Air Defense Environment 27
References 29

4. Pre–Phase A: The Air and Missile Defense Program 31
4.1 Pre–Phase A: The Air and Missile Defense Program Plan 31
4.2 Background 32
4.3 Air and Missile Defense Mission Needs and Definitions 33
4.4 Air and Missile Defense Measures of Effectiveness 34
    4.4.1 Reaction Time 35
        4.4.1.1 Engagement Timeline Definitions 37
    4.4.2 Firepower 39
    4.4.3 Defense Penetration Technique Resistance 40
        4.4.3.1 Countersurveillance and Search Phase 41
        4.4.3.2 Counterdetection and Track Phase 42
        4.4.3.3 Counterengagement and Missile Phase 43
        4.4.3.4 Counterpoint Defense Phase 44
    4.4.4 Environmental Resistance 44
    4.4.5 Continuous Availability 45
    4.4.6 Contiguous Coverage 45
4.5 Top-Level Requirements 46
References 47

5. Phase A: AMD System Requirements 49
5.1 AMD Mission Needs: Requirements to CONOPS 49
5.2 Systems Architecture Functional Requirements 50
5.3 Allocation of Functions to Systems 52
5.4 System Performance and Interface Requirements 58
    5.4.1 Central Defense System Performance Requirements 59
        5.4.1.1 Midcourse Guidance Reference Systems 61
        5.4.1.2 Handover 65
        5.4.1.3 Seeker Pointing Angle Error 68
        5.4.1.4 Midcourse Guidance 70
5.5 CDS: Sensor Suite System Performance Requirements 75
    5.5.1 Radar Architecture 82
    5.5.2 Platform Constraints 83
5.6 Engagement System Performance Requirements 84
5.7 System Requirements Document 90
References 91

6. Phase B: Preliminary Design 93
6.1 Target System 95
6.2 Sensor Suite 95
6.3 Battlespace Assessment 97

6.4			Engagement Analysis	102
6.5			Missile Subsystem Preliminary Design	107
	6.5.1		Missile Seeker Preliminary Design	107
		6.5.1.1	Angle Tracking	110
		6.5.1.2	AR Seeker Preliminary Design	112
		6.5.1.3	Signal Transmission Losses	119
		6.5.1.4	Jamming	120
	6.5.2		Translational and Attitude Response Preliminary Design	121
	6.5.3		Airframe Requirements	124
	6.5.4		Configuration Design	125
	6.5.5		Mass and Inertia Design	130
	6.5.6		Aeroprediction	131
	6.5.7		Propulsion Design	133
	6.5.8		Material Properties Design	138
	6.5.9		Attitude Response Requirements	138
		6.5.9.1	Guidance and Control Requirements	140
		6.5.9.2	Flight Control Design I	141
		6.5.9.3	Flight Control Design II	149
		6.5.9.4	Guidance Law Design	154
References				166

7. Preliminary Systems Design Trade Analysis 171
7.1 Battlespace Performance Summary 174
References 197

8. Allocation of Performance Requirements 199
8.1 Allocation of Radar Performance Requirements to Subsystems 199
8.2 Allocation of Interceptor Performance Requirements to Interceptor Subsystems 203
    8.2.1 Terminal Homing and Guidance 203
    8.2.2 Launch and Flyout Phase 204

9. Physics and Mathematics of AMD Design and Analysis 207
9.1 Interceptors and Flight Analysis 207
    9.1.1 A WGS-84 Oblate, Rotating Earth Model 207
        9.1.1.1 Transformation Matrices: Coordinate Frames and Position 208
        9.1.1.2 Transformation Matrices: Velocity and Acceleration 212
        9.1.1.3 Oblateness Effects, Nonuniform Gravity 213
        9.1.1.4 Geodetic and Geocentric Latitude Relationship 216

		9.1.1.5	Geodetic to Geocentric Latitude: Vehicle Position 217
		9.1.1.6	Latitude, Longitude, and Altitude Calculation 219
		9.1.1.7	Forces and Moments and Equations of Motion... 220

9.2 Target and Clutter Returns 225
    9.2.1 Radar Returns 225
    9.2.2 Surface Clutter Returns.. 225
    9.2.3 Volume Clutter Returns 228
        9.2.3.1 Clutter Processing Considerations 230
    9.2.4 Coherent Processing Interval 230
    9.2.5 Fill Pulses 232
    9.2.6 Maximum Clutter Range 233
    9.2.7 Clutter Rejection Degradation 233
    9.2.8 Range–Velocity Visibility.. 234
    9.2.9 Eclipsing 235
    9.2.10 Blind Velocities 235
    9.2.11 Visibility in Clutter 235
9.3 Seeker Noise Sources... 236
    9.3.1 Receiver Noise 236
        9.3.1.1 Glint.. 238
        9.3.1.2 Radome Boresight Error 239
References 240

Acronyms.. 243

Index 247

List of Figures

Figure 2.1	Three dimensions of systems engineering.	4
Figure 2.2	Pre–Phase A management plan.	7
Figure 2.3	Phase A management plan.	8
Figure 2.4	Phase B management plan.	10
Figure 2.5	Systems engineering process to SRR.	12
Figure 4.1	Pre–Phase A air and missile defense system preliminary design process.	32
Figure 4.2	Example of a ship-based missile defense timeline.	36
Figure 5.1	Assumed air and missile defense architecture.	53
Figure 5.2	Notional battlespace timeline.	57
Figure 5.3	Illustration of the keep-out volume concept.	59
Figure 5.4	Midcourse guidance and error contributions to handover.	61
Figure 5.5	CDS requirement coordinate frames.	62
Figure 5.6	Engagement segment definition.	63
Figure 5.7	Midcourse guidance terms and definitions.	64
Figure 5.8	Handover geometry.	66
Figure 5.9	Classical intercept collision triangle.	67
Figure 5.10	Missile body and seeker coordinate frame relationship.	68
Figure 5.11	Pointing angle error definitions and terms.	69
Figure 5.12	Seeker pointing angle error computational process.	71
Figure 5.13	Trajectory shaping using kappa midcourse guidance variations.	74
Figure 5.14	Kappa trajectory-shaping time-of-flight variations.	74
Figure 5.15	Process for the allocation of requirements to the radar system.	75
Figure 5.16	Notional passive phased array architecture.	77
Figure 5.17	Notional active phased array architecture.	78

Figure 5.18	Pattern propagation factor comparison of an S-band (antenna height 14 m) and X-band (antenna height 26 m) in dB under standard propagation conditions.	80
Figure 5.19	Engagement boundary requirements.	85
Figure 5.20	Engagement Mach–altitude envelope.	87
Figure 5.21	Engagement system specification development process.	88
Figure 6.1	Battlespace Engineering Assessment Tool.	94
Figure 6.2	Sensor suite specifications and preliminary design process.	96
Figure 6.3	Illustrative BMD engagement scenarios.	97
Figure 6.4	Ballistic missile defense battlespace timeline details.	99
Figure 6.5	BMD engagement target time–space correlation.	100
Figure 6.6	BMD target and interceptor time–space correlation.	100
Figure 6.7	BMD target and interceptor single-shot opportunity.	101
Figure 6.8	BMD target and interceptor multiple-shot opportunities.	101
Figure 6.9	BMD target and interceptor including discrimination.	102
Figure 6.10	Simplified interceptor block diagram.	104
Figure 6.11	(a) Target impact on terminal homing trade space for Mach 2.5 interceptor. (b) Target impact on terminal homing trade space for Mach 3.5 interceptor.	108
Figure 6.12	Phase comparison monopulse antenna representation.	111
Figure 6.13	Monopulse seeker angle tracking block diagram.	112
Figure 6.14	Narrowband MMW AR seeker versus jammer performance.	121
Figure 6.15	Wideband MMW AR seeker versus jammer performance.	122
Figure 6.16	Translational and attitude preliminary design process.	123
Figure 6.17	Notional missile configuration concept.	126
Figure 6.18	Aerodynamic trim results.	128
Figure 6.19	Representative airframe time constant.	129
Figure 6.20	Miss distance relationship to time constant.	130
Figure 6.21	Notional interceptor subsystem packaging.	131

List of Figures xi

Figure 6.22	Mass ratio calculation for specified performance.	134
Figure 6.23	Propellant load assessment.	135
Figure 6.24	Vehicle mass analysis.	136
Figure 6.25	Burn time and vacuum thrust analysis.	136
Figure 6.26	One-dimensional flyout design results.	137
Figure 6.27	Definitions of positive pitch, yaw, and roll control for $\phi = 0°$.	141
Figure 6.28	Definitions of positive pitch, yaw, and roll control for $\phi = 45°$.	142
Figure 6.29	Squeeze mode tail–fin mixing strategy.	142
Figure 6.30	Generic position servo actuator system functional block diagram.	145
Figure 6.31	Uncoupled linearized flight instrumentation measurement.	148
Figure 6.32	A simple missile flight control system architecture.	152
Figure 6.33	Simplified missile flight control system architecture.	153
Figure 6.34	Missile flight control system root locus gain sensitivity.	153
Figure 6.35	Missile flight control system closed-loop step response and gain sensitivity.	154
Figure 6.36	Missile flight control system Bode plot and margins for $K = 0.4$.	155
Figure 6.37	Missile flight control system closed-loop step response ($K = 0.4$).	155
Figure 6.38	Homing loop preliminary design model block diagram.	156
Figure 6.39	Angular noise sources and representative standard deviations.	158
Figure 6.40	Example illustrative optimal evasive maneuver weave frequency results on miss distance.	161
Figure 6.41	Lead term addition to LOS rate measurement.	165
Figure 7.1	AMD system battlespace and engagement analysis breakdown.	172
Figure 7.2	Radar detection ranges—baseline radar sensitivity and 0 dB propagation factor.	173

List of Figures

Figure 7.3	Radar detection ranges—baseline radar sensitivity and 20 dB propagation factor.	174
Figure 7.4	Notional interceptor flyout times.	175
Figure 7.5	Required detection ranges—SR ADI.	176
Figure 7.6	Required detection ranges—MR ADI.	177
Figure 7.7	Required detection ranges—LR ADI.	178
Figure 7.8	Notional battlespace chart element—one target (signature, Mach).	178
Figure 7.9	SR interceptor—5 m target, baseline radar sensitivity—0 dB propagation factor.	179
Figure 7.10	MR interceptor—5 m target, baseline radar sensitivity—0 dB propagation factor.	180
Figure 7.11	LR interceptor—5 m target, baseline radar sensitivity—0 dB propagation factor.	180
Figure 7.12	SR interceptor—10 m target, baseline radar sensitivity—0 dB propagation factor.	181
Figure 7.13	MR interceptor—10 m target, baseline radar sensitivity—0 dB propagation factor.	181
Figure 7.14	LR interceptor—10 m target, baseline radar sensitivity—0 dB propagation factor.	182
Figure 7.15	SR interceptor—50 m target, baseline radar sensitivity—0 dB propagation factor.	182
Figure 7.16	MR interceptor—50 m target, baseline radar sensitivity—0 dB propagation factor.	183
Figure 7.17	LR interceptor—50 m target, baseline radar sensitivity—0 dB propagation factor.	183
Figure 7.18	LR interceptor—5 m target, 12 dB increased radar sensitivity—20 dB propagation factor.	184
Figure 7.19	LR interceptor—10 m target, 12 dB increased radar sensitivity—20 dB propagation factor.	184
Figure 7.20	LR interceptor—50 m target, 12 dB increased radar sensitivity—20 dB propagation factor.	185
Figure 7.21	SR interceptor—5 m target, 12 dB increased radar sensitivity—40 dB propagation factor.	186
Figure 7.22	MR interceptor—5 m target, 12 dB increased radar sensitivity—40 dB propagation factor.	186

List of Figures xiii

Figure 7.23	LR interceptor—5 m target, 12 dB increased radar sensitivity—40 dB propagation factor.	187
Figure 7.24	SR interceptor—10 m target, 12 dB increased radar sensitivity—40 dB propagation factor.	187
Figure 7.25	MR interceptor—10 m target, 12 dB increased radar sensitivity—40 dB propagation factor.	188
Figure 7.26	LR interceptor—10 m target, 12 dB increased radar sensitivity—40 dB propagation factor.	188
Figure 7.27	SR interceptor—50 m target, 12 dB increased radar sensitivity—40 dB propagation factor.	189
Figure 7.28	MR interceptor—50 m target, 12 dB increased radar sensitivity—40 dB propagation factor.	189
Figure 7.29	LR interceptor—50 m target, 12 dB increased radar sensitivity—40 dB propagation factor.	190
Figure 7.30	Interceptor angle-of-attack-dependent linearized airframe characteristics.	191
Figure 7.31	Example Monte Carlo engagement results.	192
Figure 7.32	Monte Carlo engagement results summary.	193
Figure 7.33	Monte Carlo interceptor results for 7 second homing time.	195
Figure 7.34	Theoretical kill probability as a function of DOF.	195
Figure 7.35	Interceptor evaluation map.	196
Figure 7.36	AMD down selection stoplight map.	196
Figure 8.1	Radar element requirement process.	200
Figure 8.2	Number of target updates and antenna beamwidth required to support angular track accuracy.	201
Figure 8.3	Radar track time and radar frequency required to support angular track accuracy.	201
Figure 9.1	ECIC to LCIC transformation matrix rotation sequence.	209
Figure 9.2	Spherical nonrotating earth ECIC to LCIC transformation matrix.	210
Figure 9.3	Time-dependent effects due to a rotating earth.	211
Figure 9.4	ECIC rotating earth to an LCIC transformation.	212
Figure 9.5	Illustration of geodetic versus geocentric latitude.	214

Figure 9.6	Vehicle aerodynamic reference system.	220
Figure 9.7	Surface clutter geometry for a sea–mountainous land interface.	226
Figure 9.8	Illustration showing the cross-sectional view of non-beam-filled volume clutter.	229
Figure 9.9	Illustration showing the semiellipse geometry used to determine the cross-sectional area of the illuminate clutter volume.	229
Figure 9.10	Example unambiguous clutter scenario.	231
Figure 9.11	Radar returns from five coherent pulses for Table 9.1 scenario.	232
Figure 9.12	Effect of range ambiguous clutter on MTI clutter rejection capability.	234
Figure 9.13	Three types of range weighting for % coverage determination: (a) linear weighting, (b) exponential weighting, and (c) uniform/exponential weighting.	237

List of Tables

Table 5.1	AMD Architecture Functional Requirements	50
Table 5.2	Fixed Radar Design Parameters	81
Table 5.3	Summary of S-Band Passive and Active Radar Phased Array Designs	83
Table 5.4	Radar Power Requirement Summary	84
Table 6.1	Active Radar Pulse Doppler Seeker System Trade Space	114
Table 6.2	Pulse Doppler Active Radar Seeker Design Specifications and Performance	118
Table 6.3	Pulse Doppler Active Radar Seeker Acquisition Performance for Table 6.2	119
Table 6.4	Interceptor Missile Flight Regime Requirements	124
Table 6.5	Aerodynamic Model to Estimate Preliminary Configuration Properties	127
Table 6.6	Preliminary Configuration Parameters	128
Table 6.7	Interceptor Missile Weight, Balance, and Inertia Budget	132
Table 6.8	Propulsion Design Parameters	135
Table 6.9	GNC Instrumentation Strategy	146
Table 7.1	Target Radar Horizon Range Summary	173
Table 7.2	ADI Average Velocities: Minimum Intercept Ranges	175
Table 7.3	Overall AMD System Performance Summary	179
Table 7.4	Shoot–Look–Shoot Firing Doctrine Effectiveness Summary: 12 dB Increased Radar Sensitivity, 40 dB Propagation Factor	190
Table 7.5	Interceptor Zero Angle-of-Attack Linearized Airframe Characteristics	192
Table 7.6	Notional Engagement Preliminary Design Study	194

Table 9.1	Time Coincident Radar Returns for Three Pulses for Scenario	231
Table 9.2	Time Coincident Radar Returns for Four Pulses with One Fill Pulse Delay	232
Table 9.3	Time Coincident Radar Returns for Five Pulses with Two Fill Pulses Delay	233
Table 9.4	Radome Material and Characteristic Parameters	240

Preface

Balancing weapon and sensor system performance is important for achieving air and missile defense performance requirements against a target in an optimal sense. In today's environment, this means meeting performance requirements while minimizing design, development, and operational costs over the lifecycle of a combat system. This book outlines a physics-based systems engineering approach for the design and development of a balanced air and missile defense system given a fixed set of target requirements. The inherent architecture of weapon and sensor systems plays a significant role in achieving a balanced approach for negating a target in both natural and electronic attack environments.

This book represents a synthesis of knowledge that we have developed over many years of engineering experience, including interaction with a number of members from the technical community, and thus builds on the foundation of the works of many others. In this sense, the book is a progression of an evolving process that is constantly looking to new technologies and innovative architectures to ensure that air missile defense systems stay one step ahead of the target. Even as we complete this effort, the focus is shifting to networked systems of weapons and sensors to increasing air and missile defense capabilities at lower costs, which are mandated by a combination of constrained budgets and a growing competitive business environment.

We express our appreciation to several colleagues who have taken time from their busy schedules to review the draft material and provide insightful feedback. We are grateful for the comments and critique provided by Glenna Miller and Elena Zaitsev. We thank the staff at CRC Press/Taylor & Francis for their patience during the development of the material for this book.

Authors

Warren J. Boord earned a master's in mechanical engineering from the Johns Hopkins University (Baltimore, Maryland) in 1993, and a bachelor of science in aerospace engineering from West Virginia University (WVU) (Morgantown) in 1980. Upon graduation from WVU, Boord was commissioned in the U.S. Air Force (USAF) as a Second Lieutenant. After separating from the USAF in 1986 as a Captain, Boord spent the next 20 years in private industry and is currently with the U.S. Navy. Boord has more than 30 years of defense community experience, supporting the weapon systems acquisitions, scientific and technical intelligence, and military space communities. He has significant systems engineering expertise in the areas of U.S. naval surface combat systems, missile defense systems, weapon systems, satellite systems, and scientific and technical intelligence. His technical expertise includes cruise and ballistic missile defense, weapon and missile systems engineering, and missile threat systems engineering. He has technical expertise in sensors; missile guidance, estimation, navigation, and control; aerodynamics; and propulsion systems. Boord has developed a particular expertise in understanding, modeling, and simulating complex weapon systems and threats and analyzing performance results throughout the entire engagement timeline. His most recent technical publication appears in the American Institute of Aeronautics and Astronautics (AIAA) *Journal of Guidance, Control, and Dynamics* as the coauthor of a paper titled "New Approach to Guidance Law Design" (January–February 2005).

John B. Hoffman has more than 30 years of combined antenna, radar, and combat systems engineering experience. This experience includes the Aegis Combat System SPY-1 Radar, Air and Missile Defense Radar (AMDR), Cobra Judy Replacement (CJR) Radar, THAAD Radar, and G/ATOR Radar. He received a BSEE from Virginia Polytechnic Institute and State University (Blacksburg) in 1980, and an MSEE from the Johns Hopkins University in 1988. He is currently employed as a systems engineer at Systems Engineering Group, Inc. in Columbia, Maryland, supporting surface navy modeling and simulation efforts that enable shipboard combat system performance characterization in test beds and in at-sea environments.

1
Introduction and Background

1.1 Introduction

This book provides key insights into and design procedures of the air and missile defense system engineering process that results in a balanced missile defense system whose requirements are to fulfill air and ballistic missile defense needs. This missile defense systems engineering reference will provide the underlying technical foundation for missile defense engineers to conduct an organized program and analyses that will effectively guide the problem definition, investment of research and development efforts for follow-on generations of missile defense systems, and upgrades to existing missile defense systems. As such, this book will have both international and long-lasting applications.

This book focuses on shipborne missile defense systems that provide their own ship defense against missiles and protection of other nearby ships. However, the systems engineering principles discussed herein can be readily applied to other missile defense system scenarios. The goal is to provide an understanding of the physics of missile defense systems and the key performance parameters that drive the capabilities of these systems.

1.2 Why Is Missile Defense an Important Topic?

The deployment of the German V2 missile in 1944 ushered in the era of missile defense. The V2 missile provided the capability to strike targets at long ranges from mobile missile launchers. The V2 was not tactically accurate and was more of a terrorist weapon. Once the V2 was launched, there was no way to defend against it. Clearly, antimissile defensive systems needed to be developed to defend against such missiles.

In 1959, the United States placed 15 nuclear-tipped Jupiter missiles in Turkey [1]. These missiles were aimed at targets in the former Soviet Union. The former Soviet Union responded by constructing nuclear missile installations in Cuba in 1962. This all came to a head in what is referred to as the Cuban Missile Crisis. Ultimately, clear thinking prevailed and the Soviet missiles were removed from Cuba and the U.S. missiles were removed from Turkey.

Today, many countries possess both land-based and sea-based offensive missile capabilities. These offensive missiles possess accurate navigational systems based on technologies such as the global positioning system (GPS) or the global navigation satellite system (GLONASS). This allows missiles to strike stationary targets with several meters of accuracy. Advanced missile seeker designs, based on radio frequency (RF) or infrared (IR) technology, allow moving targets to be engaged with a high probability of hitting the target [2].

Clearly, the need for missile defense systems to defend against increasing and varied offensive missile system capabilities is growing. Globalization is increasing the economic ties and interdependencies between developed counties. As these economic ties and interdependencies become more entwined, the likelihood of offensive strikes among developed nations may increase or decrease. We simply do not know. Therefore, as long as offensive missile capabilities exist, missile defense systems will be needed to defend against them. In the end, missile defense may only be needed as an insurance policy against offensive missile strikes. However, it is safe to conclude that the need for missile defense will not go away.

References

1. Mango, A., *The Turks Today*, Overlook Press, New York, 2004, p. 61.
2. Bergland, E., Mission planning technology. In *Technologies for Future Precision Strike Missile Systems*, RTO-EN-018. Research and Technology Organization, Paris, France, July 2001.

2

Systems Engineering Fundamentals

There is no one theory that completely defines systems engineering (SE). Specifically, it is a branch of engineering that does not go back to physical first principles. SE is not simply a planning process to define and execute the job at hand. As the title of this book suggests, systems engineering is one piece of the total topic to be discussed. SE, however, is considered only in a loose sense, and the focus of the book is on the analysis and trade space associated with producing a balanced air and missile defense system (AMDS). SE will be treated first as an outline to accomplish the true purpose of the book. This is not a theoretical treatment of SE nor is it an exhaustive practical treatment. Accordingly, this book is not advocating for, nor arguing against, any specific SE formula. There are many examples [1–8] to choose from. It is advocated that you simply need to have one and that you try to keep it as simple as possible while not letting the details of the program slip into the cracks. Certainly, if you are spending more money on planning and executing the program than on designing, testing, deploying, and sustaining it, you have some issues to address. You should, however, be allocating about 30% of your resources to planning and requirements development.

There are more definitions for SE than could possibly fit in this book. Some have been found to be fundamentally identical and others have little in common with the rest. To avoid adding superfluous material, the definition that is most compatible with the topic of air and missile defense is adopted with some added descriptors. *Systems engineering* can be defined as "a process that is comprised of a number of activities that will assist in the definition of the requirements for a system, transform these requirements into a system through design and development efforts, and provide for the operations and sustainment of the system in its operational environment" [1]. The systems engineer is the one who is responsible for the program definition and who puts the plan of action into motion. The systems engineer has three roles: the technical roles of an architect, designer, integrator, and tester; the role of a systems or technical manager; and the role of a production engineer. To achieve success, the systems engineer is required to employ both artistic and scientific engineering skills. An experienced systems engineer develops instincts for identifying and focusing a team's efforts on activities that will ultimately achieve an optimized or balanced design while accounting for lifecycle considerations. The art of systems engineering takes the form of developing the right set of design alternatives and options and then developing the necessary trade studies that will help the systems engineer eliminate all but the best sets and combinations of alternatives from which an investment decision(s) can be made.

Why is systems engineering important? The purpose of SE is to establish a repeatable, traceable, and verifiable methodology to produce systems and products to facilitate verification, validation, and accreditation (VV&A) with an improvement in cost and schedule while minimizing the risks associated with engineering endeavors. SE includes configuration control management and lifecycle sustainability and maintainability.

Systems engineering starts by defining a standard framework within a common lifecycle process that can be applied to any system regardless of the scope or scale of the project. Numerous system engineering frameworks have been proposed. Hall [1,2] is widely quoted either implicitly or explicitly throughout the systems engineering literature [3–6]. Hall proposed that systems engineering has three major dimensions that make up the Hall morphological box of systems engineering: time, logic, and professional disciplines. Practically, this decomposition is incomplete and premature. NASA [7] proposes the morphological framework to be a three component model. Here we believe there is a fourth component and a slighted modified third component. This systems engineering framework recommended for practice is shown in Figure 2.1, which shows that systems engineering can be thought of as vectors to achieving one's goals and objectives. A program requires four component to succeed. The first component is a well thought out organizational structure or Integrated Product Team (IPT) structure. This component is the most important. Without the right organizational structure or set of structures, the program has little chance of succeeding. The second component is to populate the lead positions in the IPT (organizational structure) with experts in their field and having superior skill sets demonstrated by achievement. Third, your program needs to have established engineering standards and tradecraft practices communicated and understood throughout the team. The fourth component is to have well

FIGURE 2.1
Three dimensions of systems engineering.

established operating principals and business practices through your program. The operating principals and business process are at the center of the pyramid to communicate that the other three components alignment relies on standardized operating principals and business practices.

The concept and process essentially includes the layout of the SE plan that lends itself to VV&A and is intended to maintain consistency, repeatability, and traceability throughout the program's lifecycle. Tools and methods apply to defining tradecraft and will subsequently contribute to VV&A, configuration control, traceability, and repeatability of results. Knowledge and skills of the workforce call for the placement of highly trained and experienced engineers with appropriate backgrounds in leadership positions and, most notably, skills in running integrated product teams (IPTs). A balance between a solid government team, a contractor, and laboratory team members must also be maintained. The program should be structured such that the manager is able to matrix talent in/out of the IPTs as required.

The components of a standard systems engineering process that can be applied to any project regardless of scope and scale include the following:

- Well-defined goals and expectations (qualitative)
- Performance objectives or measures of effectiveness (quantitative)
- A concept of operations (CONOPS) that includes the way the system is intended to operate, and the way the design, test, manufacturing, and deployment process is intended to operate
- Requirements definitions that include functional, performance, and interface requirements
- Defined constraints that include itemized cost, schedule, policy, logistics, human factors, and technology
- Risk assessments that are itemized and time dependent with evolving mitigation plan
- The program's milestone objectives and lifecycle reviews

To accomplish these tasks, the SE process should be decomposed into four teams or strategies: technical management, systems architectural design, technical evaluation, and product realization. The *technical management* responsibilities include stakeholder and customer interactions; requirements, constraints, risk, configuration control management; programmatics; architectural and technical evaluation processes; planning decisions; program reviews; and the development of management documentation that includes the systems engineering management plan (SEMP). The *systems architectural design* responsibilities include requirements development and flow down; implementation of the design process and the development of the design solutions; data development; risk analysis and identification; and the development of design documentation including program review materials. The *technical evaluation* responsibilities include the development of and adherence to the systems performance analysis process; the requirements verification process; the design and end product validation process; and the development of the performance evaluation documentation packages. The *product realization* responsibilities include the formulation of the design-to-production process; the establishment of the manufacturing processes and procedures; and the development of deployment and training plans and procedures.

The next part of organizing the program should include structuring the program's milestone objectives and reviews. Each program will have to decide how many and what milestone reviews are necessary or desirable. It is important to note that reviews should not/do not end when the product is deployed.

A minimum set of timed reviews should include a systems requirements review (SRR), a preliminary design review (PDR), a critical design review (CDR), a test readiness review (TRR), an operational readiness review (ORR), an operational capability review (OCR), lifecycle assessment reviews (LAR), and a retirement and disposal review (RDR). The reviews need to be set up to include specific program accomplishments, transition decisions, and completed documentation. These reviews occur on a timeline and are embedded in a schedule. Normally, the milestone reviews are mapped to program phases that establish the entire program timeline from conception to birth to retirement (pre-cradle to grave).

The systems engineering plan encompasses the seven phases as follows:

1. Pre–Phase A: Project Definition
2. Phase A: Systems Requirements
3. Phase B: Preliminary Design
4. Phase C: Detailed Design
5. Phase D: Engineering, Manufacturing, Development; Integration and Test
6. Phase E: Operations and Sustainment
7. Phase F: Retirements and Disposal

These seven phases are defined by their purpose, activity, documentation, and the culminating program review(s). A timeline (schedule) and cost must be established and correlated with specific activities. Beyond pre–Phase A, recurring risk assessment and risk mitigation should take place. A management plan diagram can be constructed, which will identify the phase, the phase purpose, the associated primary milestone review, entry criteria associated with the phase activities, documentation requirements, and schedule requirements. Each management plan will have an associated cost and manpower requirement that needs to be secured, authorized, and funded.

2.1 Pre–Phase A

A notional pre–Phase A management plan is shown in Figure 2.2. Each of the program strategy teams is put into action starting in pre–Phase A. The management team manages this matrix, the internal processes, documentation, costs, constraints, and risks. The other teams execute their jobs as described earlier. This phase typically consists of the government team and laboratories, and should include prime contractor involvement. The prime contractors are truly the only ones who know how to engineer, manufacture, and deploy products. Your prime contractor and laboratory support teams should have already been established for planning purposes. The objective of this phase is to develop and implement the program plan. You cannot expect to accomplish practical objectives without their involvement upfront. Pre–Phase A will include defining the mission need in terms of realizable goals and objectives, concept systems and architectures, draft measures of effectiveness/performance, and systems top-level requirements (TLRs). In addition, stakeholders and their expectations are clearly identified; technology development needs are identified; and trade studies are identified and defined in detail. Iteration between these activities is expected to fully and accurately complete pre–Phase A. Verifying and validating (to be discussed in Chapter 4) results are necessary in each phase.

Entry Criteria/ Inputs	Pre–Phase A		
	Develop and Implement the Program Plan		
	Project Readiness Review		
	Activity	Documentation	Timeline
Mission need	Translate MNS into program objectives	Top-level requirements document	Months
Authorizations/ funding	Define program objective measures of effectiveness		Months
	Define top-level requirements		Months
	ID/Define constraints and risks	Systems Engineering Master Plan (SEMP)	Months
	ID/Define SE processes/methods		Months
	ID/Define knowledge, skills, abilities, requirements		Months
	Define work breakdown structure		Months
	Develop requests for proposals	RFPs	Months to a year or more

FIGURE 2.2
Pre–Phase A management plan.

The final pre–Phase A output is likely to be a project readiness review (PRR). A SEMP outline template of an SEMP is provided in Appendix 2A. The SEMP is essentially a road map to navigate through the program, which will be updated in the next two phases. Your prime contractor and laboratory support teams should have already been established for planning purposes. Prior to developing a WBS and completing the SEMP the program IPT structure must be in place. However, executing the program may require additional or different contractors. This fact will require you to send out a round of requests for proposals (RFPs) at the completion of pre–Phase A in an effort to hire the right contractor teams for the effort. The RFPs will usually be answered within a 30- to 45-day period, and the technical evaluation team will need to recommend a selection to the management team. The selection process may take months.

2.2 Phase A

A notional Phase A management plan is shown in Figure 2.3. This phase typically consists of the government team and laboratories and should, as in pre–Phase A, include prime contractors' involvement. The objective of this phase is to fully develop a baseline mission concept and assemble the systems requirements document (SRD). The final output is the systems requirements review (SRR).

Entry Criteria/ Inputs	Systems Requirement Document		
	Systems Requirements Review		
	Activity	Documentation	Timeline
Authorizations/ funding	CONOPS	Systems requirements document	Months
SEMP	Top-level systems architecture		Months
TLR	Requirements flow down		Months
SER	Configuration control management plan		Weeks to a month
	Validation, verification, and accreditation plan	Update SEMP/WBS	Weeks to a month
	Data development plan		Weeks to a month
	EMD plan		Months
	Request for proposals	RFPs	Months

FIGURE 2.3
Phase A management plan.

Phase A develops and refines feasible concepts and finalizes goals and objectives. Some concepts may be eliminated while others may be added. Systems and architectures, and measures of effectiveness/performance are refined. The TLR document is updated and approved. At this point, technology development requirements and a risk assessment are developed for each viable concept that, in turn, become part of the set of trade studies. Trade studies are executed with the purpose of eliminating bad concepts and ranking good concepts. Trade studies focus on evaluating technical, schedule, and cost objectives. The result of the trade studies includes a system and architectural baseline; functional allocations to hardware, software, and other resources; and new plans are developed. The SEMP now contains more details of the associated management plans and will be updated again in the next phase. You will be required to send out your final RFP sets. The objective here has to expeditiously send out the RFPs and then to receive, evaluate, and select contractors from the proposal responses. In some cases, the government will use the PDR in Phase B to down-select the final contract awards. In this case, the competing contractors down-selected from the pre–Phase A RFP process are funded to produce a preliminary design for review based on the documentation prepared in pre–Phase A and Phase A. Whoever came up with this concept should be considered a genius. This process allows the government to actually see the contractors in action. It provides the government with multiple design choices, which subsequently is a risk reduction activity in itself. Moreover, in the event the selected contractor slips, the government has a fallback position with the runner-up contractor. The added cost for funding multiple PDR phase competitors essentially buys down program risk while allowing multiple contractors to develop new capabilities that would benefit them and the government in follow-on competition.

2.3 Phase B

A notional Phase B management plan is shown in Figure 2.4. This phase typically consists of the government team and laboratories, and the focus is on the prime contractor developing a preliminary design for review. The objective of this phase is to produce a system definition with enough detail to baseline a design for EMD and capable of meeting the mission need. The final output is the preliminary design review (PDR). Phase B includes design and performance requirements flow down to subordinate systems and subsystems within the architecture. Interface requirements are added to the SRD. Trade studies defined in Phase A continue as required to refine concepts and are input to newly defined design studies aimed at allocating capability and performance to systems and subsystems. The design studies

Entry Criteria/ Inputs	Phase B		
	Produce System Definition Detail Necessary to Establish a Baseline Design Capable of Meeting Mission Need		
	Preliminary Design Review		
	Activity	Documentation	Timeline
Authorizations/ funding	Update SRD, flow down requirements to subsystems	Updated SRD	Months
SEMP	Place requirements under CCM		Weeks
SRD	Develop design solutions and ICDs	Systems design document	3–6 months to a year
TLR	Produce performance predictions		Months
	Develop drawings, design to specifications		Months
	Validate systems design solution against requirements	Validation assessment	Months
	Technical interchange meetings	Memoranda	Routine
	Update EMD plan	Updated EMD plan	Weeks

FIGURE 2.4
Phase B management plan.

include interfaces, which include hardware, software, and communication within the architecture, systems, and subsystems. Verification and validation plans are developed. Finally, all of the products developed in Phase A are updated and reapproved. Verification and validation accompany the results of trade studies. An updated SRD, system design documents (SDD), a verification assessment, and an updated EMD plan are presented for approval at the PDR. The SRD is now in *lockdown*. However, requirement changes beyond this point are a normal and expected occurrence. The difference is that a formal procedure and approval by configuration control management (CCM) must take place. Any proposed changes to the requirements will usually cause a cost and schedule impact to the program. If requirement changes are inevitable (requirements creep), it usually means that cost and schedule increases should be expected. Technical interchange meetings (TIMs) between all or subset IPT members are a requirement during this phase. TIMs are also highly encouraged in the earlier stages, but are shown specifically in the Phase B management plan because of the critical nature of early and continuous communication and resolution of issues between IPT members when the preliminary design is being developed.

2.4 Phases C–F

Management plans for Phases C–F are assembled in the same manner as shown earlier, but the details are different. These program phases are, however, beyond the scope of this book and are not presented. The important lesson here is that although macro schedules and plans are required, breaking your program down into small chunks is more manageable. The larger and more complex the program, the more essential it becomes to break it into smaller pieces. In turn, as you subdivide the system into smaller pieces to manage interfaces, interoperability solutions become an increasingly important design consideration and attention.

Air and missile defense systems (AMDS) are complex engineering undertakings composed of many systems within systems and subsystems within systems. AMD subsystems have many components and those components interact in complex ways with each other and the environments surrounding them. The interactions are complex and can lead to unpredictable and sometimes unexpected results. Essentially, AMD systems engineering is *rocket science* and needs to be treated accordingly.

Following good systems engineering practices will enhance the likelihood of achieving optimized performance and ultimately success in satisfying program objectives and within cost and schedule. Systems engineering employed properly will aide to produce a balanced design despite, at times, being faced with opposing and conflicting physical, fiscal, and time constraints.

Each phase discussed earlier has three common and primary focused activities that are solved iteratively. They are requirements development and management, concept development, and architectural design solution development. These activities are tied together with rigorous verification and validation activities and occur in a cyclical fashion between the activities. The iterative spiral moves from one phase to the next progressively improving the system performance.

The NASA *Systems Engineering Handbook*, SP-6105 [7], states, "The objective of systems engineering is to see to it that the system is designed, built, and operated so that it accomplishes its purpose in the most cost-effective way possible, considering performance, cost, schedule, and risk." Figure 2.5 provides a process to follow to help achieve a successful SRR that will enable these objectives being met.

Objectives and constraints are derived from the mission need and stakeholder expectations and documented. Objectives and constraints are met during each phase by defining and executing trade studies, conducting design and performance analysis, and verifying and validating results in an appropriate way for each phase. The iteration process is executed until an "optimal solution" is achieved. Optimal is the point at which the objectives

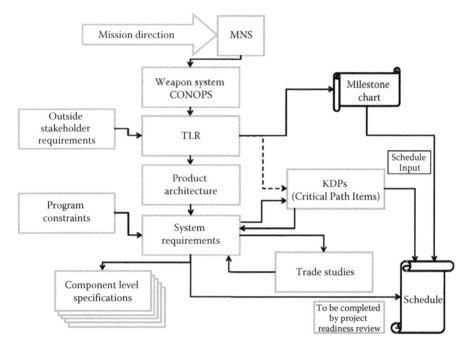

FIGURE 2.5
Systems engineering process to SRR.

and constraints are satisfied such that there is diminishing improvement in cost, schedule, and risk with additional iterations.

The three major functions that the book focuses in terms of air and missile defense systems engineering include concept-of-operations development, architecture and design development, and requirements development and analysis.

2.5 Concept-of-Operations Development

The concept of operations (CONOPS) describes the mission and how it may be accomplished. Multiple CONOPS are preferred in the beginning stages and are winnowed down to either a single or single set of CONOPS. The CONOPS is a narrative accompanied by diagrams, charts, and graphics to describe a proposed approach to solving the problem at hand. The problem is defined in terms of mission need, goals, objectives, and expectations. CONOPS is the first definition of the solution architecture. Moreover, the CONOPS will act as a tool for verification, mission planning, and requirements development and management.

Systems Engineering Fundamentals 13

Once the AMD problem is defined, the CONOPS will cover but is not limited to the following topics:

- The mission description in terms of the problem being solved
- A list of major functional requirements that need to be achieved to solve the defined problem
- A list and description of major assets considered in the solutions architecture
- A mapping of functional requirements to assets
- A list of performance metrics, both qualitatively and quantitatively, that are necessary to declare success or a level of success
- An engineering description of how the assets will perform the assigned functions including performance predictions and timelines
- A technology readiness description for achieving each functional requirement, these include technology challenges, and hurdles
- A listing and description of issues and concerns ranging from logistics, interoperability, communications, coordination, and problem identification

The CONOPS is probably not the place to address fiscal and schedule issues. It should be a source for predicting cost and schedule, which needs to be placed in the SEMP. The CONOPS needs to react iteratively to constraints brought about by schedule and cost concerns but should be independent from them to ensure that achievable technical solutions are being proposed and sought. The purpose is not to solve cost and schedule issues. It is best to not pursue the program at all if either cost or schedule prohibits a reasonable solution.

2.6 Product Architecture

Architectural design consists of engineering flow diagrams depicting functionality, interfaces, links, and communication. The product also includes graphics and technical descriptions addressing performance metrics. Once the CONOPS is produced, the objective is to produce the architecture of systems and subsystems that accomplish the CONOPS stated objectives.

Architecture is formally the style and method in which something is designed and constructed. In architecture, the structure of the components, their interrelationships, and the principles governing their evolution over the systems lifecycle are determined. The architecture is verified against the mission objective(s) and the CONOPS.

The AMDS architecture solution, most likely to evolve, is referred to as a system-of-systems (SOS) architecture or a family-of-systems (FOS) architecture. The SOS architecture according to Sage [3,4] has five principal characteristics and is composed of multiple independent systems able to achieve stated objectives on their own; the multiple systems are managed independently; they are spatially removed from one another; they develop in an evolutionary manner or spirally; and finally, the SOS has what Sage refers to as "emergent behavior." The measurement of the SOS architecture effectiveness includes measures of interoperability when defining the initial measures of effectiveness (MOE).

2.7 Requirement-Driven Acquisitions

The requirements definition process can determine the success or failure of a program. As such, it is imperative that the requirements definition processes evolve, be iterative, and involve the team at large. Moreover, the requirements must state *what* is to be achieved and not *how* it is to be achieved. Bureaucracies enjoy central planning and control, and without exception fail because they always dictate not only *what* but *how* objectives are to be achieved.

The purpose of requirements is to identify and communicate what functions are to be performed and then specify how well each component, subsystem, element, and system within the system must perform. The mission need statement (MNS) is the highest-level requirement definition and should identify a problem, the objective to be achieved, and a statement on the performance that defines mission success. The MNS is satisfied when activity produces a system that will solve the problem with the performance specified.

The design requirements are the manifestation of transforming the mission need, stakeholder expectations, and constraints into a set of executable statements used to define and develop a design solution that can be verified and validated. The design requirements communicate what functions a system must perform and how well it must perform them. These requirements are organized into a hierarchy, which flows down through the systems of interest and are typically shown in the document tree. The design requirements development process is recursively increasing in detail through each cycle and layer of the design process. Design requirements are used to develop a description of all inputs, outputs, and their relationships. Moreover, design requirements must be written in a verifiable manner.

The top-level requirements (TLR) document must include the measures of effectiveness (MOEs) for the functional assets that are required, the performance required of those assets to achieve the MOEs, and the interfaces required to achieve functionality and meet the MOEs. The TLR and the MOEs are verified against the MNS, expectations, and the program constraints. The flow down requirements are verified against the TLR and MOEs.

One additional note of importance is that constraints such as cost, schedule, and policy, among other things, are embedded in the design requirements definition process. Therefore, they are part of the trade space when defining program design requirements. Cost and schedule constraints become design performance parameters. There is no reason to develop a solution to a problem that cannot be bought, or is only available, after it is needed.

Next, understanding how "corporate" policy will affect your design or design process is essential. For example, if you are encumbered in a policy that dictates a solution to the problem you are charged with solving, it is much better to give up before you start and search for a new problem to work on. Unfortunately, policy may limit your access to resources or possibly slow down your access to resources. This in turn can cause schedule increases, will most likely drive up cost, and will possibly prohibit a more efficient, or elegant solution. These types of policy issues are usually confronted in government or in large private companies or possibly at the national level. Regardless, they need to be recognized, identified, and defined within the phases of the requirements process and dealt with early in the program.

There are three requirements components necessary to quantify the system design. These three requirements define the program and are iteratively solved during each phase of the program.

Initially, the *functional/operational requirement* is produced. Functional/operational requirements identify and define what function(s) must be performed. This activity follows the specifications provided in the MNS. For example, the MNS: *Our military forces need to destroy all hostile cruise missiles in flight before they reach their intended targets*, is a stated mission need with performance requirements that need to be satisfied. The first part of this process is to define the metrics associated with accomplishing the MNS. The metrics may also be called the cornerstones or measures of effectiveness (MOEs) of your system. The cornerstones need to define *What, When, Where,* and *How* to achieve the mission. It is also important to define *Why* the mission is to be achieved. Good examples of performance metrics are the Aegis cornerstones developed by the U.S. Navy [9]. The first part of these cornerstones, the *What*, is the defined sequence: *Detect, Control, Engage*. The *When, How*, and *Where* are listed as follows:

- Reaction time—*How*
- Firepower—*How*
- Electronic countermeasure (ECM) and environmental resistance—*When*
- Continuous availability—*When*
- Area coverage—*Where*

The cornerstones are discussed later in the book.

These critical metrics are necessary to begin the *functional requirements analysis* to identify and define the functions that need to be performed to produce a system (or a system of system—SOS) that will achieve the stated objective with the specified performance requirement. When this trade-space analysis is completed, the functions are allocated to systems in the broad sense. Constraints and risks are identified and defined, and the CONOPS is produced, documented, and verified against the MNS and the specified performance objective.

The next requirement set needed is the *design performance requirements*. This is achieved by a trade-space analysis that is referred to as the *systems performance requirements analysis*. During this analysis process, an engineering balancing act is performed to distribute the responsibility for achieving the MOEs derived from your MNS objective with the specified performance using the systems where the required functions were allocated. Once completed, a systems requirements document (SRD) is developed, where the associated constraints and risks are also identified and maintained.

The next fundamental sets of requirements to develop are the *interface and integration requirements*. Interface requirements begin to develop during the PDR Phase B by conducting the appropriate trade-space analysis. System alternatives are selected in this design trade-space analysis where integration across systems is analyzed as part of the approach to determine the appropriate systems required to reach performance, cost, and schedule objectives. Once completed, an *architectural design* can be developed, and the associated constraints and risks are identified and documented. It is in the interface and integration requirements development process that interoperability issues are identified and derived requirements are developed to address those issues.

Each of these requirements components has associated constraints that must accompany their development. The requirements bundle is developed in a hierarchy flow decomposed into the three main components and become more detailed as the design and development process progress. Shared requirements across elements may exist and will be stated within the requirements breakdown wherever they exist. In Phase A, the systems requirements document (SRD) contains all three requirement types. In Phase B, the requirements have been flowed down into detailed specifications that allow a detailed design that satisfies the Phase A requirements to be built in Phase C.

The SRD will contain the top-level requirements and also specify the mission success criteria. The natural requirements flow will have the performance requirements following each functional requirement. Interface and integration requirements then specify how the elements interoperate by specifying their connectivity and are followed by interface performance requirements. The performance requirements have to be written and communicated in a verifiable manner.

2.8 Verification and Validation

Verification and validation are accomplished during each cycle of the design and development and for each segment, system, subsystem, element, and component to ensure the system meets the required mission objectives.

Essentially, verification addresses whether or not the design satisfies the requirements. Each of the major system elements, CONOPS, and the architectural design are verified against the requirements and must be consistent in solving the objective problem. Verification continues throughout the design and development process in a sequential manner. CONOPS is verified against the mission objective; the architectural design is verified against the CONOPS.

Validation is usually defined by ensuring that the objective system is built correctly. This includes analyzing, inspecting and testing, and simulating the system prototype against real-world data to accomplish validation. In the case of an AMDS, it should be tested and simulated in a manner that represents the way it is intended to fight. This requires end-to-end testing and simulating, including all input items, interfaces, and performance requirement emulations. Expected and unexpected variations in the environment and intended target sets need to be explored. Often, it is not possible to accomplish such extensive testing in pure hardware. Simulations become more important when the system and the problem become more sophisticated. In many cases, simulation may be the only means to validate the design that leads to an entirely new program of modeling and simulation that should be handled in a parallel manner to the systems development and testing against real-world data whenever possible. To build and test complex and sophisticated simulations has now become an inherent requirement for AMDS engineering. This not only includes a digital representation of your system but a commensurately complex digital representation of the intended targets and the associated environments. This is no small undertaking, and it will be a necessary part of your budget.

2A Appendix: Systems Engineering Management Plan

2A.1 Background

The systems engineering effort begins with the pre–Phase A concept study clearly identifying and communicating the program objectives. The systems engineering management plan (SEMP) is the foundation document communicating technical, engineering, and management activities for the program. The SEMP is generated during milestone Phase A, and baselined in Phase B. The SEMP should be updated as necessary when major project changes occur. The details of schedule, work flow, and the order of activities should be continuously updated as part of ongoing planning. An SEMP is prepared to address the complete milestone execution of the program. More specifically, the purpose of the SEMP is to communicate the technical and management road map by addressing technical design and development approach, process, integration, test and evaluation, and interfaces. SEMP documentation and approval serves as an agreement within the program of how the technical work will be conducted. The program manager usually assigns the SEMP responsibility to the systems engineer or the technical equivalent. The technical IPT leads work with the systems engineer and program manager to contribute, review the content, and obtain concurrences as necessary.

2A.2 SEMP Outline

2A.2.1 Introduction

The objective of systems engineering is to see that the system is designed, built, and operated so that it accomplishes the intended mission need and expectations in the most cost-effective way possible, considering performance, cost, schedule, and risk. The SEMP provides an end-to-end road map to accomplish this objective.

2A.2.1.1 Program Purpose

The program purpose is contained in the Mission Needs Statement (MNS). The MNS is a stated mission need with performance requirements that need to be satisfied.

2A.2.1.2 Program Mission Overview

Clearly, one should describe and document the mission objectives to ensure that the program team is working toward a common goal. The program objectives and any newly evolved objectives form the basis for performing the mission, and they need to be clearly defined and articulated. The program constraints, appropriate to the mission, are also captured and used to verify the mission design.

2A.2.1.3 System Concept-of-Operations Overview

The CONOPS is initially developed as a draft concept during pre–Phase A, with refinement throughout the lifecycle, until the flight operations plan is completed in Phase D. The outcome and decisions for key operations concept trade studies and optimizations should be documented. Trade studies and analyses are used to demonstrate that the operations concept will meet the mission requirements including cost and schedule and is consistent with the architecture and design. The CONOPS is verified against the TLR document including mission objectives. The CONOPS should be used to conduct requirements identification, flow down, and management, leading to the requirements used to generate the architecture and design.

2A.3 Integrated Product Team Structure and Responsibilities

2A.3.1 Technical Management

Define the systems engineering organizational chart with roles and responsibilities. The program manager should assign a chief or lead system engineer. The program manager and the systems engineer should develop the plan for the systems engineering effort and establish a system engineering team along with roles and responsibilities. This plan, along with the roles and responsibilities, are captured in this section. The systems engineer coordinates the efforts of the systems engineering team and has the responsibility for the systems engineering functions and products for the overall program. Specific duties and responsibilities are delegated by the systems engineer with the program managers' approval to other members of the system engineering team and are captured here. Communication protocols between the team and the technical management are established here. Coordination protocol requirements are established here. The program manager makes decisions that maintain a balance between technical and programmatic performance.

2A.3.2 The Development of the Architectural Design Solution

The architectural design solution is first generated in pre–Phase A and defined and refined until the end of Phase B at the PDR. Initially, the architecture should start out as functional or logical blocks. As the design matures, the architecture should mirror the physical product breakdown structure (PBS). Once block and flow diagrams and interfaces are defined, then detailed design (Phase C) can proceed, with minimal risk of a major change induced by an architectural block diagram change. The architectural design solution should decompose the total system into its major parts to form the hierarchy for lower-level interfaces and specifications. The major parts of a system include the separate subsystems and boxes and their embedded hardware and software functions. The major goal of AMD systems engineering is coordinating the engineering trade studies, design, and development of an architectural solution that meets the requirements, is consistent with the CONOPS, operates in the mission environment, and can be developed on schedule and within cost. The SEMP should identify the technical means by which the architectural design solution will evolve and be documented throughout the team. Typically, block and signal flow diagrams are the primary means for documenting and communicating the architectural solution and designs to the team.

2A.3.3 Product Development

Production typically involves a different set of IPTs compared to any of the previous development phases. These IPTs as well as their technical management structure, and roles and responsibilities need to be defined.

2A.4 The Milestone Review Process

The project lifecycle is defined as a set of phases: formulation, approval, and implementation. The SEMP defines systems engineering phases that can be defined as pre–Phase A, Phase A, Phase B, Phase C/D, and Phase E/F terminology. Each systems engineering phase consists of functions and a workflow, which produce the products needed for the completion of the phase. The mission review is the verifying event for the phase and results in a revised mission baseline. The lifecycle accommodates the objective of systems engineering by considering implementation alternatives in Phase A, completing a preliminary design and verifying that the right system has been designed in Phase B, performing a detailed design and verifying that the system is designed correctly in Phase C, building and validating the system in Phase D, and operating and disposing of it in Phases E and F.

2A.5 Requirements Identification and Analysis Process

Requirements communicate what functions a system must perform and how well it must perform them. They describe the interfaces a system must meet. Requirements should be organized into a hierarchy that flows down through the systems of interest. The levels of requirements are typically shown in a document tree. The mission level-1 requirements, usually defined in the project plan, define mission success criteria and serve as the top level for the requirements hierarchy. Some of the items that need to be defined in the SEMP are listed in Sections 2A.5.1 and 2A.5.2.

2A.5.1 Requirements Identification

Document the requirements appropriate to the complexity of the system element. Define who develops the requirements hierarchy and who is responsible for each part of the hierarchy. Define what format is planned and what tools if any are used for documenting and tracking the requirements. Define when requirements identification is due and when formal configuration control is expected to start. Functional requirements describe what the system must do. Describe how to identify functional requirements. Performance requirements are attached underneath their respective functional requirement. Performance requirements describe and document how well the function needs to be performed. Performance requirements are written in a verifiable manner. Describe how to identify performance requirements. Requirements should specify the interfaces or reference configured interface specifications. Interfaces and ICD requirements are planned interfaces to be included. Describe how to identify interface requirements.

2A.5.2 Requirements Management

Requirements may be organized into functional, performance, and interface categories. The requirements flow hierarchy should be consistent with the product breakdown structure. Define who is responsible for developing the ICDs and who has approval and configuration management authority. Requirements are decomposed and allocated to products down through the PBS. Ideally, this continues until a single engineer is responsible for the product. Some shared requirements may flow between and across subsystem elements. Define who identifies and is responsible for the crossover requirements.

2A.6 The Verification and Validation Process

The SEMP should provide the systems vision for verification and validation (V&V). V&V are interrelated and are accomplished throughout the systems engineering process and include lifecycle support and sustainment. Together, V&V demonstrate that the AMD system meets the mission need, design goals, and stakeholder expectations. This is referred to as the mission objectives.

2A.6.1 Verification

The SEMP should describe how verification is used to ensure that the AMD design satisfies the mission objectives. It is a continuing process that encompasses the verification of CONOPS, the architectural design, and the requirements. The CONOPS, requirements, and architectural design verification process should ensure that mutual consistency between them is maintained throughout the program. The written process should emphasize that Phase A and B verification activities strive to show that the right system design has been chosen before the detailed design proceeds in Phase C, minimizing the chance that the wrong system will be designed. Verification also occurs later in the lifecycle when mission simulations, end-to-end tests, and other activities show that the AMD system has been designed correctly and meets the mission objectives.

2A.6.2 Validation

Validation is an important risk reduction function that attempts to uncover issues before they become operational problems. Validation includes those functions, which ensure that the team builds the system correctly, by validating all requirements and verifying the architectural design against the requirements. The validation process includes the identification of the item and the method (analysis, inspection, or test) for validation. The process for the review and approval of the validation results needs to be explained. The validation activities of Phases C and D show that the correct system is designed.

2A.7 Configuration Control Management Process

Configuration control management (CCM) is a library for documentation, software, and designs; provides version definition and control; and facilitates access and dissemination. Products are placed into the CCM to serve as a single point of reference for the program. The SEMP should reference a CCM plan that identifies the CCM officer and specifies the products necessary for inclusion in CCM, and the version control process associated with each type of product. A process for the identification and use of latest revisions is required. Validation is dependent on a robust CCM process.

2A.8 The Risk Management Process

The SEMP should define who is responsible for risk management and provide the references for risk management assessments. Within the SEMP, define the risk assessment philosophy and what risk analyses are planned, who is responsible, and how the analyses are to be accomplished, including any special tools. Define when and how often risk analysis is due.

References

1. Hall, A.D., *A Methodology for Systems Engineering*, Van Nostrand, Princeton, NJ, 1962.
2. Hall, A.D., A three dimensional morphology of systems engineering, *IEEE G-SSC Transactions*, 5(2), 156–160, April 1969.
3. Sage, A.P., *Handbook of Systems Engineering and Management*, Wiley Series in Systems Engineering, John Wiley & Sons, New York, 1999.
4. Sage, A.P., *Methodology for Large-Scale Systems*, McGraw-Hill, New York, 1977.
5. Kossiakoff, A. and Sweet, W.N., *Systems Engineering Principles and Practice*, Wiley Series in Systems Engineering, John Wiley & Sons, Hoboken, NJ, 2003.
6. NASA/SP-2007-6105 Rev 1, *Systems Engineering Handbook*, National Aeronautics and Space Administration, Washington, DC, December 2007.
7. Thomas, D., Systems engineering guideline development overview, NASA Presentation, September 15, 2005.
8. Hill, J.D. and Warfield, J.N., Unified program planning, *IEEE Transactions on Systems, Man, and Cybernetics*, 2(5), 610–662, November 1972.
9. Global Security.org, Aegis program history. http://www.globalsecurity.org/military/systems/ship/systems/aegis-history.htm. Accessed on January 9, 2016.

3
Missile Defense Problem

3.1 Overview of the Missile Defense Problem

A missile defense system essentially has four functional requirements that combine to provide defense against an incoming missile. These elements require balancing and, as such, this presents the missile defense systems with engineering challenges. These elements are intelligence, surveillance, and reconnaissance (ISR); detection and tracking; weapons control; and engagement. Ballistic missile defense is divided into three parts: boost, midcourse, and terminal intercept engagement phases. Cruise missile defense is typically divided into three components: area, self-, and point defense phases. Self-defense and point defense phases may utilize the same system components but have different mission requirements. These elements and missions combine to provide a layered defense capability to maximize defense performance.

One of the most effective defense penetration techniques is to collapse the battlespace by minimizing the engagement time available. The primary techniques available to collapse the battlespace for the offensive missile designer to exploit are [1] speed, altitude, and radar cross section (RCS). The defense system in turn must utilize faster missiles, elevated and more powerful ISR sensors, and radars with data links and sophisticated signal processing techniques to counter these offensive techniques.

After collapsing the battlespace, the offensive missile designer needs to drive down the probability-of-kill (P_k) or probability-of-raid annihilation (PRA) for the defensive systems that have obtained an engagement opportunity [2].

Offensive missile tactics and raids can be used to reduce P_k or PRA. Tactics can include jamming and maneuvers either in combination or separately [1–9]. Jamming is employed to delay detection by the radar and missile seeker and to deny the radar and missile seeker accurate range and angle estimates.

Evasive maneuvers are one of the most, if not the most, effective tactics used to evade defensive weapons such as missile and radar-directed gun weapon systems [7–9] and bring down P_k.

Raids are used to saturate and confuse the defensive systems and can be stream or simultaneous [3]. Stream raids are a series of missiles on the same trajectory with some time spacing between the individual missiles. Simultaneous raids are designed such that all offensive missiles arrive at the target almost simultaneously. An example of a simultaneous raid is an azimuth raid where the offensive missiles are separated in azimuth angle but arrive concurrently [3].

The offensive missile speed is a defense penetration fulcrum that will synergistically add to the weapon's ability to collapse the battlespace and reduce the P_k/PRA of a defensive system. For example, if an inbound offensive missile is traveling at Mach 1 and is detected by the radar at a slant range of 56 km, when can it be engaged? The system reaction or latency time determines how quickly a missile can be launched against an offensive missile once it is detected. If the system reaction time is 10 seconds, the offensive missile will travel 3.43 km or 1.85 nautical miles (nm) in 10 seconds to a slant range of 52.13 km or 28.15 nm. If the missile used to engage the offensive missile also flies at Mach 1, the offensive missile can be engaged at 26.08 km or 14.08 nm after a 76 second flyout time. If the weapon system can fire a second shot 5 seconds later, a second engagement attempt can occur at 25.21 km or 13.61 nm in 2.5 seconds after the first attempt. The flyout time of the second missile is 73.5 seconds.

If the offensive missile is inbound at Mach 3 and detected at the same range of 56 km or 30 nm, how much is the engagement timeline reduced for the same missile? Now, the offensive missile will travel 10.28 km or 5.55 nm in 10 seconds to a slant range of 42.28 km or 24.45 nm. A Mach 1 missile can intercept the offensive missile after a flyout time of 32.95 seconds at a slant range of 11.30 km or 6.10 nm. The 3-to-1-speed mismatch between the offensive missile and the missile reduces the range of the first engagement opportunity by approximately 14.82–11.30 km (8–6.1 nm). The second engagement opportunity occurs at 10.0 km or 5.40 nm in 1.2 seconds after the first attempt. The flyout time of the second missile is 29.16 seconds.

Finally, if the offensive missile is inbound at Mach 1 and a Mach 3 missile interceptor is employed, how does the scenario change? Now, the offensive missile is at a slant range of 52.13 or 28.15 nm when the first missile is launched. The offensive missile can be engaged at a slant range of 39.08 km or 21.10 nm after a missile flyout time of 38 seconds. Now, the 3-to-1-speed mismatch increases the first intercept range by 13 km or 7 nm relative to the first example. The second engagement opportunity occurs at 37.78 km or 20.4 nm in 3.8 seconds after the first attempt. The flyout time of the second missile is 36.76 seconds.

Clearly, mismatches in speed can be used to the advantage of the defensive or offensive weapon. The key is to stay technologically ahead of the offensive missile by developing higher speed weapons. The ultimate weapon in terms of speed is a directed energy weapon (DEW). DEWs, such as a high-power laser or a high-power microwave beam, travel at the speed of light. Of course, these types of futuristic weapons need to be very accurately directed particularly at longer engagement ranges. Moreover, one of the most important objectives in air and missile defense is to place as much energy on the target as far away from the asset being defended as possible. The energy penalty on DEW with respect to range is a steep cliff. In addition, adverse weather conditions such as fog or rain will likely reduce the lethality of DEWs in all but the shortest ranges.

The offensive missile can also use altitude to its advantage by flying below the radar horizon as long as possible [4,5]. This denies the radar detection of the offensive missile until it is visible above the horizon. The range at which the line of sight between the radar and the offensive missile becomes unobstructed is related to the height of the radar and offensive missile. At lower altitudes, the maximum unobstructed range can be approximated by

$$R(nm) = 1\ xx(h_{radar} + height_{offensive\ missile})^{1/2}$$

where the number "1 xx" is an approximation that has to consider earth location and environmental factors such as whether you are over water or land. Clearly, the height of the radar must be increased to extend the radar horizon. If the radar height is increased by a factor of 4, the radar physical horizon can be doubled for low-flying offensive missiles such as "wave skimmers" or terrain following cruise missiles. For land-based and shipborne radars, the ability to increase radar height is limited. Land-based radars can be placed on elevated terrain (e.g., hilltops) although this makes them more conspicuous to the enemy. Shipborne radar can be placed high on the superstructure or mast. The size and weight of the radar also need to be considered. In general, smaller and lighter weight radars can be placed at higher locations on a ship. Another consideration for the placement of ship-borne radars is placing the radar antenna(s) to minimize obstruction zones, particularly at lower elevation angles.

Radar cross-section reduction can be used to delay radar detection. In clear environments, the radar detection range is proportional to $1/r^4$ where r is radar slant ranges. This means that the radar must be 16 times (12 dB) more powerful to double the detection range against a fixed radar cross-sectional target. Radar range can be increased through a combination of increased antenna gain (larger antennas or higher frequency for fixed antenna area), increased transmitter power, or reduced losses. Active phased array radars reduce losses compared to passive phased radars by reducing both transmit and receive losses. This is accomplished with the use of transmit/receive (T/R) modules in a distributed architecture. Signal processing techniques can also be used to increase detection range. Noncoherent or coherent integration across multiple pulses can be employed to increase detection range at the expense of slower search frame times. Concentrating radar resources in the offensive missile sector can also be used to increase detection range.

Offensive missiles can reduce their radar cross section by employing shaping or radar-absorbing material (RAM) [4]. Surface shaping causes the energy in the transmitted radar pulse to be reflected in other directions and not directly back to the radar, causing the energy in the radar return to be reduced. The offensive missile can also use composite materials that absorb radar energy to reduce their radar cross section. Other techniques related to the offensive missile seeker involve tilting the seeker when it is not active and the use of a seeker radome constructed of a frequency selective surface (FSS). Tilting the seeker will reduce the in-band forward aspect radar cross section. In-band refers to the offensive missile seeker band, which may be different from the air defense radar band. This is essentially a simple shaping technique that directs most of the radar energy at bistatic angles to minimize the monostatic radar cross section. The use of an FSS for the seeker radome material reduces the forward aspect radar cross section out of the offensive missile seeker band. The FSS is essentially transparent in the seeker band and reflective out of band. The out-of-band forward aspect RCS is reduced by the radome shape.

Stream and azimuth raids are employed to improve defense penetration capability [2,3]. Both raid types will place increased demands on the defensive systems' firepower capability in an attempt to overwhelm the air defense system. The defensive systems can potentially counter these techniques with increased magazine capacity, increasing the number of engagements that can be prosecuted simultaneously and by reducing the salvo time (how rapidly missiles can be sequentially fired from a launcher).

A stream raid is defined as a raid where a group of similar offensive missiles fly the same trajectory, but the trajectories are separated in time [2,3]. Modern radars should have sufficient range resolution to detect and track each offensive missile in the raid. However, sometimes, modern radar may ignore trailing targets that occur in the same beam or at the same indicated angle. An offensive missile can generate false trailing targets if it employs a repeater jammer with digital radio frequency memory (DRFM) [4–6]. The transmitted radar pulse is digitized, stored, and then retransmitted at fixed time delays to mimic false trailing targets. A Doppler shift can be added to the retransmitted pulse that gives the false trailing targets the same apparent inbound velocity as the actual offensive missile.

If the radar identifies each missile in the stream raid as an offensive missile, then special consideration should be given to the engagement solution. If possible, trajectories for the weapons used to engage the trailing offensive missiles should be shaped to avoid debris from the engagement of leading offensive missiles. Trajectory shaping will increase missile flyout times. Stream raid engagements will become more difficult for smaller stream raid time spacings and faster offensive missiles.

An azimuth raid is a raid where the offensive missiles fly toward a common target from different azimuth angles. The raid timing is such that all of the offensive missiles will arrive simultaneously. This scenario is judged to be a worst-case scenario from an engagement timeline perspective. Most defensive systems do not have the ability to engage multiple offensive missiles in azimuth simultaneously. This is due to several reasons. All offensive missiles in the raid will not be detected simultaneously due to radar search patterns and the stochastic nature of the detection process. Usually, weapons can only be launched one at a time at an interval determined by the missile launcher salvo time.

3.2 Air Defense Environment

The environment can and will have significant performance impacts on a defense system. The radar system and RF missile seekers have to contend with multipath, clutter, and jamming environments individually and in combination [2–6]. Multipath signals can interfere with direct path signals causing cancellation or fading for some geometries. In addition, multipath causes errors in the offensive missile elevation angle estimate, resulting in the uncertainty of the offensive missile's true altitude. Clutter effectively raises the noise floor of the radar making target detection much more difficult without clutter cancellation. Noise jamming also raises the radar noise floor, reducing target detectability. The jamming power at the radar is reduced proportionally to $1/r^2$ where r is the range between the jammer and the radar. The signal level of the radar return has a $1/r^4$ since a two-way path is involved between the radar and the offensive missile (transmits and receives). This relationship gives the jamming platform the ability to standoff at significant distances and still be effective. Other types of jamming, particularly jammers on board the offensive missile, try to deceive the radar and/or missile seekers range and/or angle estimates. These techniques are referred to as deceptive jamming.

The most benign environment is a clear environment, which is defined as a smooth earth surface and standard propagation conditions. In this environment, radar performance is limited by the radar horizon and multipath from the smooth earth surface. Multipath is an indirect path from the offensive missile to radar that involves a reflection from the earth's surface. The indirect path is longer than the direct path, which results in a phase difference in the radar signals arriving via the direct and indirect paths. There is also a small time delay in the indirect path relative to the direct path due to the increased distance traveled by the radar signal. When the phase difference due to the combination of the path length differences and phase of reflection coefficient at the earth's surface is 180°, the radar return is essentially canceled. At the other times of the phase, direct and indirect path radar returns can be in the phase, resulting in a 6 dB enhancement in target signal-to-noise ratio relative to the direct path alone. Radars or missile seekers that have wide operational bandwidths can adjust their transmit frequency as a function of offensive missile range to avoid multipath cancellation. Approximately 30% operational bandwidth is required. Otherwise offensive missile tracks can potentially be coasted during range intervals where multipath nulls occur.

Surface clutter returns, such as those from land or sea clutter, result when the surface area is illuminated by the radar beam. The illuminated area is a function of range from the radar and is bounded by the range by the pulse width and azimuth or cross-range by the azimuth beam width. The contributions from the surface clutter in this illumination region integrate or combine to determine the received clutter-to-noise ratio, which is a function of range and the mean reflectivity of the illuminated clutter patch.

Volume clutter results from rain or chaff in the radar beam. The volume is bounded by the pulse width in range and the antenna beam width in azimuth and elevation. The contributions of the clutter in this volume combine to determine the received clutter-to-noise ratio, which is a function of the range squared and the mean reflectivity of the illuminated clutter volume in m^2 per m^3.

The clutter returns have a Doppler response. Land clutter is stationary and does not produce a Doppler response. Vegetation on the land such as grasses and trees will sway in the wind and produce small Doppler responses, which have some spectral spread. The Doppler component and spread of sea clutter is determined by the wind and sea state, which are interdependent. Whether the clutter is viewed from downwind, upwind, or crosswind directions also influence the Doppler components. Volume clutter, which is airborne, is also influenced by the wind speed and turbulence. Wind speed generally increases with altitude, therefore the Doppler component from volume clutter returns increases with the altitude of the clutter.

Signal processing techniques are used to cancel clutter. These techniques are based on the Doppler difference between the clutter and the target. Clutter typically has a low Doppler response compared to a high speed inbound offensive missile. Common clutter cancellation techniques are moving target indicator (MTI) and pulse Doppler (PD). Both techniques require coherent processing of multi-pulse dwells and are ultimately limited by a pulse-to-pulse and intrapulse phase and amplitude instabilities in the radar system hardware.

Wideband noise or barrage noise jammers generally try to cover the radar operational bandwidth. For any given radar pulse or dwell, the instantaneous bandwidth used by the radar is only a fraction of the operational bandwidth. The noise in instantaneous bandwidth of the radar reduces the signal-to-noise ratio of the radar return from a target or offensive missile. If the noise jamming raises the radar noise floor by 10 dB, the signal-to-noise ratio will be reduced by 10 dB relative to a clear environment.

The jamming affects the return through the receiver antenna patterns sidelobes or mainlobe. Sidelobe jamming can potentially be canceled or mitigated by adaptive placing nulls in the receive sidelobes in the jammer direction. Sidelobe jamming is automatically reduced by how far the sidelobes are below the main beam peak. For low sidelobe phased arrays, the receive sidelobes are typically 40–50 dB below the main beam peak. Sidelobe jamming cancellation becomes more difficult as the number of jammers increases, as the instantaneous bandwidth of the radar increases, and when the jamming occurs in the main beam.

Main beam jamming provides the greatest challenge to the radar designer. Brute force radar signal processing techniques sometimes referred to as *burnthrough* can be used to form skin tracks on main beam jammers. These techniques require multi-pulse integration to overcome the jamming. Burnthrough techniques can overcome main beam jamming at the expense of radar resources. Other signal processing techniques exist to provide main beam jamming cancellation when there is some angular separation between the target and the jammer. These techniques essentially place a main beam null in the jammer direction while maintaining the gain of the main beam in the target direction.

In real-world scenarios, radars and missile seekers must be able to operate in complex environments that include the combined effects of multipath, clutter, and jamming [2–6]. Missile seekers must also contend with terrain bounce jamming (TBJ) and towed decoys. Both of these deceptive techniques are designed to confuse the seeker angle estimates for the target [2–6]. TBJ is designed to make the seeker fly into the ground instead of engaging the actual target [5]. The towed decoy is designed to capture the seeker and have it attack the trailing decoy versus the actual target [2–6].

References

1. McEachron, J.F., Subsonic and supersonic antiship missiles: An effectiveness and utility comparison, *Navy Engineers Journal*, 109(1), 57–73, 1997.
2. Polk, J., McCants, T., and Grabarek, R., Ship self-defense performance assessment methodology, *Naval Engineers Journal*, 106(3), 208–219, May 1994.
3. Phillips, C., Terminal homing performance of semiactive missiles against multitarget raids, *AIAA Journal Guidance, Control and Dynamics*, 18(6), 1427–1433, November–December 1995.
4. Bergland, E., Mission planning technology. In *Technologies for Future Precision Strike Missile Systems*, RTO EN 018. NATO Science and Technology Organization, Paris, France, July 2001.
5. Schleher, C.D., *Introduction to Electronic Warfare*, Artech House, Dedham, MA, 1986.
6. Jing, Y.-T., A survey of radar ECM/ECCM, *IEEE Transactions on Aerospace and Electronic Systems*, 31(3), 1110, July 1995.
7. Zarchan, P., *Tactical and Strategic Missile Guidance*, 3rd edn., Vol. 176, Progress in Astronautics and Aeronautics, AIAA, Washington, DC, 1997.

8. Lipman, Y., Shiner, J., and Oshman, Y., Stochastic analysis of the interception of maneuvering antisurface missiles, *AIAA Journal Guidance, Control and Dynamics*, 20(4), 707–714, July–August 1997.
9. Yanushevsky, R., *Modern Missile Guidance*, CRC Press, Taylor & Francis Group, Boca Raton, FL, 2008.

4

Pre–Phase A: The Air and Missile Defense Program

4.1 Pre–Phase A: The Air and Missile Defense Program Plan

This chapter deals with establishing the fundamental knowledge necessary to understand the elements of air and missile defense (AMD) systems engineering and the initiation of the pre–Phase A planning stage. From a technical perspective, which this book is addressing, there is more than enough material to simply lay the foundation for what the air and missile defense mission need is such that the purpose of this book can be fulfilled. There are no new discoveries presented in this chapter. There are no new revelations for the casual reader of daily newspapers, and TV news program viewers are not already exposed to what is being presented here. The references for this chapter [1–6] will provide the reader sufficient background material to explore in more detail the global expansion of missile weaponry that is no longer contained in the technologically sophisticated centers of the United States, Western Europe, and the former Soviet Union. This book does not address proliferation, which is the particular entity responsible for the spread of *first-world* weaponry, or treaties intended to reduce proliferation, but simply acknowledges that proliferation has happened and will likely continue to happen. Without any interest in the political interpretation of the reader, the authors acknowledge that the world owes President Ronald Reagan the credit, gratitude, and admiration for his visionary accomplishment of putting in motion the notion that it is necessary, accomplishable, and admirable to establish defenses against missile attacks.

To this end, the remaining sections of this chapter will follow an abridged outline of Figure 2.2 and present the development of the air and missile defense mission need statement and translate it into a technical programmatic plan. The program objectives are established in terms of achievable measures of effectiveness (MOEs). A top-level set of requirements will then be developed. The remaining pieces defined in Figure 2.2 are left to the practitioner.

4.2 Background

The AMD-specific system's pre–Phase A engineering process is developed and evolved in this chapter. Figure 4.1 shows the proposed multifaceted process flow that incorporates the main body of accomplishments flowing down the center of the diagram and establishes disciplined activity flow to produce the preliminary design and the outline for the remainder of this book.

Recursive arrows to the right of the diagram depict that verification and validation (V&V) and risk analyses are an integral part of the design process. V&V are not left for the final stages of the program activity but are essential to the entire design and development process. Another way of looking at it is that system validation and verification are designed into the system and are not a check at the end. Risk is continuously evaluated, and the results of this analysis are incorporated into design decisions and process activity. V&V and risk analysis techniques and methods are usually specific to agencies and organizations. Techniques and methods are found in the references provided in Chapter 1. The NASA *Systems Engineering*

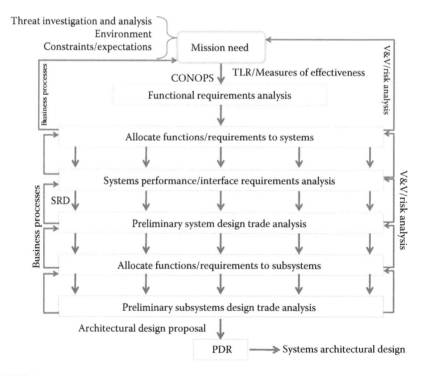

FIGURE 4.1
Pre–Phase A air and missile defense system preliminary design process.

Handbook (chapter 1, reference 6) has a mature and executable set of V&V and risk processes.

The recursive arrows on the left-hand side of Figure 4.1 indicate the business processes that must occur. Again, these processes are embedded in the design process and include matters involved with funding your program(s), executing contracts and communicating milestone progress, delays, and hardships. The business activity area is where you account for identifying programmatic and organizational constraints and ensuring that they are communicated to systems engineering.

The mission need statement (MNS) arises from a multitude of occurrences. As Figure 4.1 indicates, a threat is identified through investigations and analysis. The threat is qualified and quantified in both geopolitical and technical terms. The current political and military environment will dictate policy impacting solutions and approaches to accomplishing the mission. The expectations and associated constraints combine with the need to negate the threat to form an MNS.

From this point, the process is product oriented. The products will include a top-level requirements (TLR) document, the concept of operations (CONOPS), system requirements document (SRD), and the architecture design proposal (ADP). These products will be produced according to the schedule of the phases defined in Chapter 1, along with the other products defined for each phase.

Top-level requirements are part of the pre–Phase A planning and will flow from the MNS. As part of the TLR development process, the measures of effectiveness are developed to initiate a functional requirements analysis that will lead to the SRD.

4.3 Air and Missile Defense Mission Needs and Definitions

An MNS might read as follows: *There exists a requirement for free and peaceful nations to render offensive missile weapons useless for the purposes of deterrence, to enhance power projection ability, and to produce military advantage when force is necessary.* Thus, the MNS establishes a need to construct an air and missile defense system. It would continue that the AMD system(s) must be cost-effective, reliable, and readily available. In addition, this system or set of systems must be upgradeable and responsive in a timely manner.

It is necessary to digress to truly understand the meaning of cost-effective. When discussing defensive systems, cost-effective has often been mischaracterized as the cost ratio of expended engagement asset to offensive missile asset. This is simply not the metric that defines the trade space accurately. This comparison epitomizes the old saying—*This is like comparing apples to oranges.* The accurate cost-effectiveness trade space in engineering terms is the cost ratio of expended engagement asset to the protected asset. This is an important distinction. The costs for the user nation to build, deploy, and expend the inbound missile are not a relevant factor. There is no correlation between how much it costs one nation to build a particular missile in terms of both real dollars and with respect to their gross domestic product and another nation to build and deploy the exact same missile and more importantly the AMD system necessary to defeat it. There is, however, a technical, military, and monetary correlation between how much it costs a particular nation to build an asset (ship, port, power plant, etc.) that needs to be protected and the cost for an AMD system to protect it. For example, it can be said that it is too costly to defend a specific asset, but not that it is too costly to shoot down a threatening missile. This is another important distinction that needs to be made. Take a specific example of ship air defense. If each air defense missile (ADM) expended costs $1M to achieve a theoretical probability of kill (P_k) of 0.99 in self-defense of a $2 billion ship against a hypothetical cruise missile worth $500K to the belligerent nation, three ADMs must be expended. The question is "Does the ship fire only one missile to maintain a reasonable cost ratio?" For this example, the cost ratio is not 2:1 in favor of the defender; it is 4,000,000:1 in favor of the offense. Of course, this is not the computation that needs to be made. The actual approach to determining cost-effectiveness would be to calculate the cost of three expended ADMs to the cost (or more importantly value) of the protected asset, which is a 0.00015 ratio showing conclusively a very cost-effective system. Of course, this does not even put a value on the lives at stake or the consequences of the outcome. These calculations are left for military analysts and are not within the scope of an engineering treatment of the problem. What is important here is to immediately dismiss cost arguments without correlated elements. Defense will surely cost more than offense; however it is the cost and value of the defended asset that is correlated with the cost of its defense.

4.4 Air and Missile Defense Measures of Effectiveness

The term *measures of effectiveness* implies achieving a specific set of results. Returning to the example in Chapter 2, it is stated that the first part of this process should define the metrics associated with accomplishing the MNS or program charter and that the metrics may also be called the cornerstones of your system. The cornerstones need to define *What*, *When*, *Where*, and *How* to achieve the mission. The MNS defines *Why* the mission is to be achieved. The Aegis cornerstones developed by the U.S. Navy [7] are an example of these measures. To properly achieve the desired war-fighting capability in the defense against missile attack, one must examine the intelligence, surveillance, and reconnaissance (ISR); detection and tracking; weapons control; and engagement processes in the context of the entire defense system mission areas defined earlier. The *When, How,* and *Where* are listed as follows:

- Reaction time—*How*
- Firepower—*How*
- Defense penetration technique resistance—*When*
- Environmental resistance—*When*
- Continuous availability—*When*
- Contiguous coverage in theater—*Where*

Each of these metrics is defined in detail within the context of the overall objectives in what follows.

4.4.1 Reaction Time

Reaction time refers to executing a specific weapon system defense strategy successfully by ensuring that all operations occur within the available engagement timeline. Depending on the mission, as discussed earlier, the engagement timeline requirements will change and the flow down of requirements on the entire system will need to be reevaluated. It is therefore important to lay down the various events associated with the engagement problem. The specific event timing will change depending on the specifics of the engagement problem. For example, a short-range ballistic missile (SRBM) will achieve burnout, apogee, and impact sooner than a medium-range ballistic missile (MRBM). This, in turn, will require detection and engagement activities to occur sooner for engaging the SRBM. Therefore, first, recognize that there is a target timeline and then develop a defensive weapons system that supports target negation within the engagement timeline. The engagement solution involves careful consideration of both.

The target timeline can be broken down into two major target sets: ballistic targets (targets whose flight path extends outside of the sensible atmosphere and returns through the atmosphere to engage its targets) and air targets (referring to targets that fly strictly within the sensible atmosphere). First, the ballistic missile defense event timeline is examined, as shown in Figure 4.2.

As the title suggests, Figure 4.2 illustrates a ship-based ballistic missile defense (BMD) engagement timeline but can be generically applied to BMD from any platform. The target has a boost, midcourse, and terminal phase of flight. Within these phases, and not depicted in the graphic, are other possible important events that need to be timed depending on the target. For example, multiple-stage missiles will have staging/separation

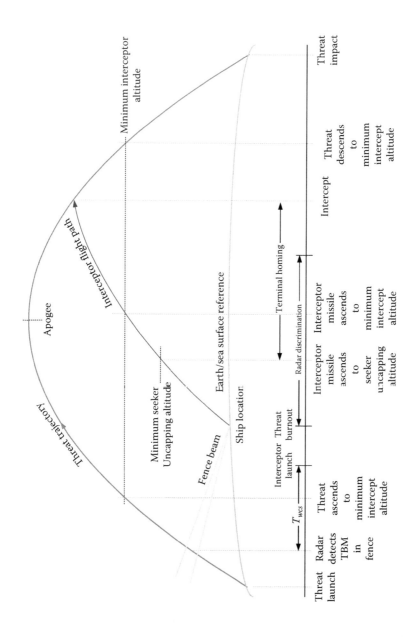

FIGURE 4.2
Example of a ship-based missile defense timeline.

events. Most modern ballistic missiles will have attitude control, thrust termination, and other events that may be important to missile defense. The event timing and the details of the events are important to the engagement timeline and must be examined in detail. Separations, terminations, other types of events, and lethal object discrimination requirements can and will contribute to the complexity of the engagement.

The intersection of the target timeline with the defensive system timeline is dependent on the engagement strategy being employed. Figure 4.2 shows, in general, the events that need to be accomplished to consummate a ballistic missile engagement. More detailed engagement events including the weapons fire-control solution, missile discrimination, and other important processes are not discretely depicted. However, Figure 4.2 demonstrates the complex nature of the engagement and that the defensive system must capture the target as early as possible, manage resources efficiently, and engage as early as possible to prosecute an engagement. In this example, it is shown that a detection fence is built with the intent that if the target flies through the fence, it will be detected. This can and most likely will be accomplished with the aid of inorganic assets as part of the ISR process not shown in this diagram. The timeline between detection and missile away will be referred to as the defense system time constant. It is within this period that a number of system functions must take place. The functions will likely have to include computing a fire-control solution (intercept capability) and challenging the validity of the potential target, normally referred to as identification friend or foe (IFF). It is not possible to completely generalize all AMD system architectures and therefore leaves open the possibility that other critical functions may be required that will utilize time.

4.4.1.1 Engagement Timeline Definitions

The following formal definitions apply for the purposes of this book. System reaction time—sometimes referred to as the defense system time constant shown in Chapter 5—is the time from initial target detection to first missile motion. System reaction time consists of three major components: (1) target detection and transfer of target location from sensor to shooter, (2) combat ID, and (3) missile launcher response time. These times can be stochastic in nature due to performance variations associated with the equipment and operators (for man-in-the-loop systems). The time required to transfer target information to the shooter is critical in systems that require lock-on before launch (LOBL). These systems are typically fire-and-forget systems, which no longer require target state data from the sensor that initially detected the target once the missile is launched. These types of missile systems can have their own seeker—active RF seeker, passive IR, or dual-mode systems. Usually, dual-mode systems are both active or passive RF and passive IR seekers. Some missile launchers use a laser to guide the missile to the target.

These systems must either have the laser slaved to the sensor data or acquire the target on their own to maintain laser guide on the target. Both passive IR seeker and laser-guided systems may be severely degraded when operating in fog or rain environments.

The second function that contributes to the overall system reaction time is combat ID. This can be performed at the unit level or in a centralized location. Unit-level combat ID is usually automated and based on both radar data and transponder interrogation to provide identification friend or foe (IFF) data from the combat system. A target at a high inbound speed, which does not respond to IFF interrogations, would normally be classified as hostile. Unit-level combat ID usually takes less time than centralized combat ID. The centralized combat ID typically requires the sensor target data to be transferred via a communication link to the centralized command center. The target information will be reviewed and may be correlated with data from other sensors in order to determine if the contact is threatening. Once the decision is made that the contact is hostile, contact information needs to be relayed back to the firing unit over a command link in order to initiate the engagement process.

Missile initialization: Once a specific missile is chosen for engagement, the third contributor to the engagement timeline includes enabling the missile to be launched and performing in-flight functions. Specifically, pyrotechnic squibs are usually used to initiate battery operation, begin the warhead arming sequence, as well as initiate other functions. Inertial platform initiation may include spinning up gyroscopes unless nonmechanical systems are used such as ring laser gyroscopes. The launcher system also has a response time, which is the time between *pushing* the launch button and the first missile motion. The launcher response time can include lock-on time for LOBL systems and launcher slew time for systems that need to slew in azimuth and/or elevation. For missiles that are lock-on after launch (LOAL) or are midcourse radar guided via missile uplink commands, launcher response times are generally shorter. LOAL systems with vertical launch cells are preferred and require a minimum amount of launcher preparation time.

Doctrine: Another dimension of the battlespace timeline is the doctrine employed. Doctrine refers to the preplanned approach for engaging targets that can be tailored to mission-specific needs. The doctrine will include the approach to the allocation of resources. More specifically, radar resource management strategy is included in the doctrine. This includes rules and conditions for rolling back radar resources in order to ensure that the high-priority engagement functions always have enough resources to support successful engagements. The weapon employment strategy includes planning the number and type of weapon or weapon variant(s) to expend in a specific engagement opportunity. This also includes consideration of the weapon magazine load out and what weapons remain in the magazine at any given time.

Salvo time: The time interval between successive launches in a specific engagement opportunity is referred to as salvo time. Firing doctrines that include multiple salvos, such as Shoot–Shoot (SS) and Shoot–Shoot–Shoot (SSS), are used to increase overall kill probability; however, this is at the expense of depleting the missiles in the magazine more rapidly. Salvo time generally includes launcher timing limitations and sufficient timing between missile shots to avoid missile fratricide. Missile fratricide can occur when the second shot in the salvo guides to the first missile in the salvo instead of the intended target. Scheduling sufficient salvo time between successive shots is one technique used to prevent fratricide.

In conclusion, all of these times and their management are requirement drivers that must be considered in developing a system-level engagement solution. The requirements' flow down process will capture these drivers within the appropriate elements to ensure a balanced systems approach.

4.4.2 Firepower

Firepower refers to having the ability to place ordnance on the target when and where they are needed with sufficient numbers to ensure success. There are two firepower requirements that need to be addressed. The first component is being able to reach the target with a sufficient amount of range at intercept between the inbound target and the defended asset that its survivability is ensured. The second component is a homing requirement. The ordnance must also be able to achieve a successful miss distance to achieve a *kill* or achieve the desired single-shot probability of kill (P_{ssk}). A flow-back requirement results from the homing requirement. It is also necessary to reach the target with a sufficient number of missiles as possible to achieve the necessary P_{ssk} to ensure the destruction of the target, which will guarantee defended asset survivability. Not only must a sufficient number of ordnance rounds reach their target, but also the right ordnance must reach the target. Not all missiles are created equal. To ensure homing success, choosing the missile to reach and the design to handle any peculiarities of the target must be ensured. The combination of reaching the target at sufficient range with the correct and sufficient number of ordnance and achieving homing success should establish firepower requirements.

The firepower requirement then contributes to defining doctrine requirements mentioned in the previous section. Firepower is heavily dependent on the amount of timeline available for the engagement(s). Any engagement may and probably will require multiple weapons (ordnance) to be placed on the target to achieve an acceptable success criterion. Each weapon expended will require revisiting the battlespace timeline at least in part. This sequence of expending weapons is also part of the doctrine. This part of the doctrine is referred to as the firepower doctrine. A firepower doctrine requires establishing a defense strategy for successfully negating all incoming targets.

Depending on the mission, other design decisions, and the tactical situation, the firepower doctrine requirements may change and the flow down of requirements on the entire system will then need to be reevaluated.

Five specific firepower doctrine strategies for this book are defined: Shoot; Shoot–Shoot, Shoot–Shoot–Shoot, Shoot–Look–Shoot, and Shoot–Shoot–Look–Shoot. Each term, shoot, refers to the launching or expending of a weapon (ordnance) to intercept the incoming target. Each time a weapon is expended, it takes battlespace time. Each term, look, refers to a kill evaluation and the cycling to a new and independent shot.

A specific doctrine is chosen to produce a kill while expending a minimum amount of resources and maximizing the amount of resources available to begin a new and independent engagement. The doctrine should also include the policy used to select the *correct* engagement option and firing doctrine to employ. Firepower doctrine requirements are developed based on the P_{ssk}.

The firing doctrine then establishes essential timeline requirements driving the weapon system time constant requirements. And conversely, the firing doctrine will be driven by the achievable weapon system time constant. This iteration will eventually settle when achievable combinations of engagement solutions, time constants, and firing doctrines are found that will result in the required P_{ssk} for the defined target set overall P_k requirement. This solution set will define firepower.

The required firepower will likely only be accomplished with a layered or tiered approach to AMD. Considering the possible incoming target design variations, including cruise and ballistic missiles and aircraft (manned and unmanned), covering the extent of the atmosphere and beyond, it is unreasonable to expect that deploying one weapon system will handle all possible engagements. Missile speeds ranging from low to high subsonic and supersonic through hypersonic (Mach 5 and above) all need to be engaged thus placing different and competing requirements on the AMD system. The trade space for achieving the firepower objectives is complicated and will involve evaluating different weapon concepts and firepower doctrines. These trades are beyond the scope of this book.

4.4.3 Defense Penetration Technique Resistance

Defense penetration techniques (DPTs) are defined as the design measures employed by the potentially hostile adversary in their offensive air and missile systems that are intended to defeat the defensive systems defending those assets desired to be destroyed. As discussed earlier, cruise or ballistic missile defense is attempted by one of three means. Either a hard or soft kill solution is employed or in combination. The adversarial offensive missile design team can break up the problem into four generic flight phases to design counters necessary to defeat the entire system. This set will be referred to here as the *time-phased defense penetration design options*. These options are designated as follows:

1. Countersurveillance and search phase
2. Counterdetection and track phase
3. Counterengagement and missile phase
4. Counterpoint defense phase

It is based on the adversary design approach that resistance to these measures and techniques is found. Performance requirements cannot be established until each of these time-phased defense penetration options (DPOs) are evaluated and a set of them is selected to be addressed in the AMD system design.

Each of the four *time-phased defense penetration design options* will be described in detail in this section. In adversary studies, the AMD system and a defense penetration design approach are decided upon. The options will consider the strategy necessary to defeat the kill chain by most likely spreading out the challenges across each phase. Speed is the defense penetration fulcrum providing the primary leverage feature dictated by physics for all penetration phases offering advantage to the offensive missile designer. All other defense penetration design measures and techniques become more effective the faster the offensive missile travels. To compress the engagement timeline (collapse the battlespace), the adversary may employ high speed and low radar cross section, high speed and low altitude, or any of a large number of combinations of defense penetration techniques, where speed notably reduces the signature reduction and altitude lowering requirements for the offense [8–15].

4.4.3.1 Countersurveillance and Search Phase

The first step in the kill chain usually begins with surveillance and search. Within this phase, potentially hostile systems or vehicles are determined to exist or not. When detected, an assessment of their potential hostile intentions or activity is made through the battle management process. If determined to be of a hostile nature, a battle management plan is invoked and appropriate action is taken to transition and designate it as a target and to begin an engagement. This engagement solution will provide the essential time needed for executing the remaining pieces of the kill chain resulting in a successful engagement. As with all the phases, time is the most critical resource to have or deny depending on which side of the problem you are on. Surveillance and search can be accomplished from space-, air-, sea-, sub-surface-, and/or ground-borne assets. These assets can be either organic or inorganic to the actual shooter or defense system.

The objective of the adversarial missile design team is to deny, degrade, or confuse surveillance systems to buy time. It is only reasonable to assume that eventually, the target missile will be found and an engagement process will begin. *Countersurveillance and search design options* include organic, non-organic, and networked sensor assets. These sensor assets may be from a single system (organic), external systems (inorganic), or a diverse highly interconnected set of platforms (network). Design options may include stealthy features, concealment CONOPS, deceptive trajectories, and various forms of electronic countermeasures. The mission during this phase of the attack is typically a cat-and-mouse game where each side attempts to understand the likely CONOPS and capabilities of the other. The AMD system requirements include developing methods and techniques to defeat countersurveillance and search techniques.

4.4.3.2 Counterdetection and Track Phase

The AMD system that is protecting a given air space from missile attack will ultimately need to detect and transition to track the target within a fire-control system. The fire-control system and sensor do not necessarily have to be co-located but are part of a system used to conduct the engagement. The search and track sensor(s) may be, and usually are, part of the same system and provide the transition between the first and second *time-phased defense penetration design option* phases. The distinguishing part of this phase is that it starts after the missile is on its way and heading for a particular target and the search sensor has detected the target. The detection may occur from autonomous search or as a result of a cue from external (inorganic) surveillance systems.

The objective of the target design during this phase is to *reduce reaction time* by employing various design options aimed at the AMD system. This may include trajectory and altitude variations, jamming, signature reduction, masking schemes, and speed [8–14]. These *time-phased defense penetration design option*s (which will be referred to as DPO from here on) will be employed in a cost-effective manner, which is most advantageous to penetrating a particular AMD system. This means that the DPO design and employment strategy will be dependent on the AMD system it must defeat and penetrate. The DPOs will be employed singularly and in combinations as the task dictates. It is important to remember that the target design need only *reduce reaction time* and not deny detection to be successful in this phase. The question the offensive missile designer needs to ask and answer is how much time reduction is enough. The answer to this question depends on the counter-DPO capabilities of the defensive AMD system. The objective of the AMD system design is to limit the effectiveness of target DPO through performance trade-offs that will result in robust counter-DPO design requirements.

The longer the time an offensive missile is successful in denying detection and track, the simpler and less difficult it is to defeat the engagement and missile phase. The AMD requirements challenge is to develop design measures that effectively resist DPO aimed at reducing reaction time. Techniques that can increase system reaction time include the use of elevated sensors to extend the radar horizon and increased radar sensitivity to mitigate the effects of target RCS reduction. Techniques that can maintain performance in reduced system reaction time scenarios are faster missiles to reduce the flyout time and missiles/ordnance with capability to engage at very short ranges. These techniques and methods available to the AMD system designer must be itemized, prioritized, and characterized before achievable requirements can be developed.

4.4.3.3 Counterengagement and Missile Phase

Once the missile target is detected and begins to be tracked, an AMD weapon system fire-control or engagement solution is computed. The time it takes to produce a solution and conduct all of the necessary checks and schedule the other necessary events needed to support the engagement and produce first missile motion is named the weapon system time constant (T_{WCS}), as shown in Figure 4.2. Simply said, this is the time it takes from detection to missile away. It is within this process that the system computes the likely point in time and space where the engagement is likely to end. This spatial location is called the predicted intercept point (PIP), and the time is called the estimated time to go (TGO). The accuracy of these two parameters is dependent on the precision, accuracy (quality), and resolution of the sensor track information and the computational approach employed to resolve PIP and TGO from the measured data. Subsequently, the accuracy of PIP and TGO will determine the accuracy of the missile midcourse guidance commands used to guide the missile during this phase of flight. A perfect midcourse guidance law will only guarantee that the missile will go where it is being sent. However, it will not be sent to the correct location if all of the supporting data and computations are also not perfect. Imperfect guidance laws, noisy sensor measurements of the target state, and computational inaccuracies contribute to the radar-to-missile handover error. Handover error, simply stated, is defined by the fact that the missile is not placed in space and time at the end of midcourse guidance that would require the missile no additional effort to intercept the target in the remaining engagement time. Homing time is the portion of the engagement where an onboard missile sensor detects and tracks the target and completes the engagement by providing measured data to an onboard guidance computer where acceleration commands are produced and executed through a flight control system. It is during the terminal homing that the handover errors are to be removed. If not, the missile will miss the target and the offensive missile will ingress to the next phase of the engagement.

The objective of the target design team is to explicitly defeat this part of the system and kill chain. To accomplish this, the target design characteristics must be able to induce unacceptable missile miss distances so as not to be effectively negated or to deny computation of an acceptable fire-control solution. The ways to produce such miss distances are numerous. The methods can start by attacking the quality of the track through deceptive countermeasures or by countering the missile in flight through maneuvers or jamming techniques [9–15]. Fire control and doctrine denial can be accomplished by simply flying so fast through this phase that a successful fire-control or doctrine solution is unachievable.

The counterengagement and missile phase may be the weakest or most vulnerable link in the engagement chain and may be where most of the defense penetration design options will be employed [9–15]. It is therefore incumbent on the AMD designer set to focus on requirements that minimize system susceptibility to these methods and techniques. Fortunately, physics limits both sides of this problem equally and solutions can be found for the AMD system. Again, the techniques and methods available to the offensive designer must be itemized, prioritized, and characterized before achievable AMD requirements can be developed.

4.4.3.4 Counterpoint Defense Phase

Short-range systems will be employed by the defended asset to add a layer of self-defense for *leakers*. Leakers are targets that have successfully penetrated the outer layers of the defensive system (the engagement and missile phase of the kill chain). Short-range self-defense systems can include missiles, guns, and electronic countermeasure techniques [9]. Most of the weapons employed here are simply fire-and-forget systems that do not rely heavily on sophisticated fire-control systems for in-flight guidance.

The objective of the offensive missile design team is to explicitly defeat this part of the system and kill chain as in the previous phase. Again, the way to accomplish this is to induce unacceptable miss distances so as to not incur damage or destruction. The ways to produce such miss distances or render jamming countermeasure systems ineffective are numerous. Trajectory variations and maneuvers are proven means to significantly reduce P_{ssk} and penetrate the last layer of defense [13–15]. Utilizing dual-mode guidance systems to avoid homing in-bands that are being jammed has also been shown to be effective [11–13]. As in the countermissile phase, the AMD requirement here is to focus on minimizing system susceptibility to these methods and techniques. The techniques and methods available to the offensive designer must be itemized, prioritized, and characterized before achievable requirements can be developed.

4.4.4 Environmental Resistance

Environmental resistance establishes the requirement to maintain AMD performance in adverse environments. Originally, this was to include jamming environments. It is believed that all of the methods and techniques employed by the offensive system(s) to penetrate defensive systems are more effectively captured by measures that resist the target defense penetration techniques (DPTs), which are discussed in the next section. This allows the systems engineer to develop design requirements that focus on operating and fighting in degrading weather and propagation environments against a single measure of effectiveness (MOE). The environment itself as it affects electromagnetic (EM) and electro-optical (EO) propagation must be met with performance requirements that ensure a graceful degradation of the AMD system engagement performance. In fact, these will evolve as interface requirements for the entire AMD system. Some of the sources for environmental degradation include atmospheric absorption including rain attenuation of the EM and EO spectrums; rain-induced backscatter clutter; sea clutter return; atmospheric refractivity including anomalous propagation (ducting); atmospheric property variations including density, pressure, and temperatures; low-altitude tracking errors including multipath and lobe cutting; background radiation (most prevalent in the EO/IR environment) interferences; and target signature anomalies such as glint and scintillation.

The AMD system requirements must be developed to accommodate reduced performance realities as these environmental effects are analyzed. The good news is that both sides of the fight will need to operate in the same environment and will suffer degraded performance. It is the design that has best accounted for environmental variations that will have the advantage.

4.4.5 Continuous Availability

Continuous availability refers to the elements or components of the AMD system solution that must be operable at all times during deployment. Operable has to be defined as operating with the expected probability of achieving design objectives. Otherwise, the remaining requirement definitions that hinge on achieving a specified P_{ssk} will not be achievable. This book will not dwell or elaborate on this MOE or the requirements it produces as it is not within the desired scope of this book. This falls on the art and engineering of manufacturing and reliability. It is important to this MOE that redundancy is built into the system at critical nodes so as not to field a system with single points of failure or an *Achilles heel*.

4.4.6 Contiguous Coverage

A reference frame has to be created before it is possible to establish a contiguous coverage MOE. The reference frame can take on many definitions but must include seamlessly enveloping the defended asset with protection from all azimuths out to a specified range and altitude. This is likely accomplished with tiered protection having overlapping engagement area responsibilities. In the surface warfare example, a battle group must be configured to provide protection of the carriers, LHAs, and possibly civilian assets without gaps in coverage. This can be accomplished with multiple air defense–capable ships working under a specific CONOPS for the situation. These *fighting ships* may have different capabilities relative to various target sets and range-azimuth sectors. The ships must therefore protect themselves in order to be available to contribute to the AMD system capability. Thus, the AMD system must be required to operate independently as well as part of a larger, more encompassing capability. Attaining and maintaining the contiguous coverage MOE will levy requirements on the AMD system that the *elements* be interoperable, therefore creating an additional set of interface requirements [9].

4.5 Top-Level Requirements

The proper beginning of the top-level requirements is to state or restate the mission need statement. This would not be done here. The TLR must first address the fundamental mission objective. The mission objectives develop the technical facts associated with accomplishing the objective of the MNS and with the MOEs used as constraints and assumptions.

The next task is to clearly define the problem. The problem is the target(s). Target set engineering characterizations that include flight dynamic envelope and sensor correlated signature details are the essential elements necessary to establish defensive requirements. Moreover, it is necessary to characterize target physical attributes that will affect lethality requirements; employment options that may include coordinated attacks, reattacks, and waypoint usage and varying speed and altitude regimes that may affect timeline requirements; and the active use of deceptive and interference techniques such as electronic countermeasures (ECMs) [11–13]. These characterizations will then need to be mapped into the MOEs.

Performance and interoperability requirements will be formed by identifying the regions of the world where the AMD is to effectively operate, identifying the need for interoperating with different forces and assets, and identifying specific environmental conditions. The TLR includes operational engagement altitude-range regimes, probability of asset survivability requirements, single-shot probability of kill (P_{ssk}) requirements for defensive missile solutions, and time-on-target probability of kill requirements for energy weapons and guns. Raid densities will be incorporated with all P_{ssk} requirements.

The functional performance and interface requirements will evolve from a mapping of these considerations and the target characterizations into the MOEs. The TLR document product will then be used to flow down to the concept of operation (CONOPS) and the development of the system requirements document (SRD).

References

1. Gromley, D.M., The risks and challenges of a cruise missile tipping point, Monterey Institute for Nonproliferation Studies, Monterey Institute for International Studies, Monterey, CA, September 2008.
2. Reports to Congress Pursuant to the FY 2000 National Defense Authorization Act, Annual Report(s) on the Military Power of the Peoples Republic of China.
3. Woodside, R.E., Feasibility of Third World advanced ballistic and cruise missile threat, in *Presented to 67th MORS Symposium Working Group 4: Air and Missile Defense*, Washington, DC, June 24, 1999.
4. Tissue, P. et al., Attacking the cruise missile threat, Joint Forces Staff College Joint and Combined Warfighting School Class #03-3, September 8, 2003.
5. The threat, http://missilethreat.com/thethreat/pageID.133/default.asp.
6. The endgame, *Inside Missile Defense*, January 7, 2004.
7. Global Security.org, Aegis program history. http://www.globalsecurity.org/military/systems/ship/systems/aegis-history.htm.
8. McEachron, J.F., Subsonic and supersonic antiship missiles: An effectiveness and utility comparison, *Navy Engineers Journal*, 109, 57–73, 1997.
9. Polk, J., McCants, T., and Grabarek, R., Ship self-defense performance assessment methodology, *Naval Engineers Journal*, 106, 208–219, May 1994.
10. Phillips, C., Terminal homing performance of semiactive missiles against multitarget raids, *AIAA Journal Guidance, Control and Dynamics*, 18(6), 1427, November–December 1995.
11. Bergland, E., Mission planning technology. In *Technologies for Future Precision Strike Missile Systems*, RTO-EN-018. NATO Science and Technology Organization, Paris, France, July 2001.
12. Schleher, C.D., *Introduction to Electronic Warfare*, Artech House, Dedham, MA, 1986.
13. Maksimov, M.V., *Electronic Homing Systems*, Artech House, Norwood, MA, 1988.
14. Lipman, Y., Shiner, J., and Oshman, Y., Stochastic analysis of the interception of maneuvering antisurface missiles, *AIAA Journal Guidance, Control and Dynamics*, 20(4), 707–714, July–August 1997.
15. Yanushevsky, R., *Modern Missile Guidance*, CRC Press, Taylor & Francis Group, Boca Raton, FL, 2008.

5
Phase A: AMD System Requirements

5.1 AMD Mission Needs: Requirements to CONOPS

The air and missile defense (AMD) problem [1–7] requires capability to engage advanced aerodynamic and ballistic missiles, aircraft, and unmanned autonomous vehicles (UAVs) in a complex tactical theater. This theater will involve the operation of air, sea, and land platforms from various service organizations and possibly from coalition forces. Air and missile defense (AMD) will include theater defense, area defense, point defense, and self-defense. The AMD mission will pursue engaging the threat at the earliest opportunity utilizing combined theater assets.

Theater defense can be defined as a war-fighting asset that provides protection to any other asset throughout the theater of military operations. The protected entity can be a military or civil set of assets or population centers. Area defense can be defined as the protection of those military assets within a combat war-fighting grouping. An example of area defense is the Aegis area air defense, antiair warfare (AAW) mission [1–3]. This example includes the requirements for the war-fighting ship (Aegis) to defend the carriers and all other ships within a battle group against attack by an air threat. Point defense is the mission to protect assets from within the immediate region of the missile attack. This can be the protection of military or civil assets or population centers. The patriot advanced capability (PAC-2 or PAC-3) or patriot air defense missile system [6] is an example of a point defense system. Point defense also includes self-defense. When combinations of these strategies are employed, it constitutes a layered defense system where the systems operate in succession to eliminate the threat. Referring back to Chapter 2, these systems acting together may be called a family of systems (FOS) and, as such, suffices as our concept of operation (CONOPS).

The program objective is to develop a *war-fighting system* to defeat the target set before it can either achieve its mission by design or inadvertently achieve a mission success by disabling or destroying either intended or unintended assets of interest to those defending against the attack.

49

5.2 Systems Architecture Functional Requirements

The architecture is defined based on the functional requirements that will flow from the CONOPS. Five functional requirements are identified to satisfy the CONOPS as described in Section 5.1 [1,3,4]. These functional systems are a central defense system (CDS), an Intelligence, Surveillance, and Reconnaissance (ISR) system, a target system, an engagement system, and a communication link system (CLS). These functional requirements and the functions they are required to perform are shown in Table 5.1.

The central defense system (CDS) is defined as the organic elements that prosecute the engagement. The ISR system consists of those elements that produce target cueing information and pass that information to the CDS for action. The target system is the set of targets and support systems that need to be destroyed or dismissed through discrimination and identification processes. The engagement system consists of the set of weapons available to destroy the lethal segment of the target system. The CLS is the set of communication channels that provide interoperability between all of the elements.

A flow down of requirements for each functional element is shown in Table 5.1. The CDS is shown to have five functional requirements. It is required to: (1) detect, track, and discriminate the target set; (2) compute the engagement doctrine and develop a set of decisions that will dictate the sequence of engagement events based on the target track data and AMD system performance data and engagement predictions; (3) iteratively update engagement solutions that are based on stored engagement system performance capabilities and the target track data and produce a predicted intercept point that is passed on to the engagement system and updated; (4) illuminate the target

TABLE 5.1

AMD Architecture Functional Requirements

Central Defense System	ISR System	Target System	Engagement System	Communication Link System
Detect and track, discriminate	Compute target origin	Signatures	Flyout	Frequency
Compute doctrine and decisions	Compute and update target geoposition	Dynamics	Midcourse guidance and control	Bandwidth
Compute engagement solution/PIP	Predict target geospatial/temporal end goal	Time/space correlation	Terminal homing	Word content and format
Illuminate/handover	Communicate	Physical attributes	Lethality mechanism	Data rate
Communicate			Communicate	Reference frames (spatial, temporal)

Phase A: AMD System Requirements

(as necessary for semi-active radar systems) and/or point the engagement element sensors to the target at an established handover point in time and space; and (5) establish communication data links and passageways to each affected system and element in the proposed architecture.

The ISR system has four distinct functional requirements; it must compute the target origin; compute and update the target geolocation; predict the target time and space end point; and establish communication data links and passage ways to the proposed CDS.

The target system characteristics will drive the AMD architecture and design features. Target signatures are those associated with the operating bands of the sensors chosen within the sensor suite system and within the identified scenarios. Target dynamic bounds must be characterized, and the time and space correlation of the dynamic bounds and signatures must be produced. In addition, the physical attributes that will affect lethality decisions and performance will be required. In a capability-based acquisition, these target parameter characterization spaces will be used to bound and select new capabilities to be acquired for each spiral or upgrade that increases AMD capability.

The engagement system(s) will have five functional requirements. The engagement system weapon will need to fly out to intercept the lethal target set; guide (possibly navigate) under radar control during the midcourse phase of its flight; transition to a terminal mode for precise homing to intercept the target; employ a means for lethal termination of the target; and, throughout this entire process, communicate to the other interdependent elements within the AMD architecture.

Midcourse guidance includes the requirement to fly the engagement system to the PIP computed by the engagement computer system (ECS) and communicated through an uplink containing high data rate acceleration commands and have the engagement system produce a transponder signal and downlink with position and kinematical data. Terminal homing can be passive, active, semi-active, or multimode. Lethality mechanisms are fragmentation warheads and hit-to-kill (HTK) kinetic energy systems.

The CDS may also possibly pass engagement system skin track data and produce illuminator functions if semi-active radar (SAR)-guided missiles are involved or pass continuous homing commands as is the case in a track-via-missile (TVM) architecture. Each of the communication and illumination links will have a data rate and data bandwidth requirement associated with the transmission. Data links need to be robust to provide acceptable performance in electronic attack (EA) environments.

Of course, this set of engagement system functional requirements sounds suspiciously like a missile. Most likely, missiles will be the primary solution for any type of near- or midterm AMD system but possibly not exclusively. Besides the likely integrated use of electronic warfare (EW) systems, guns as an engagement system will also fit into the functional flow described here. High-energy weapons (HEWs) are a possible far-term option to integrate into the engagement set, and they must also satisfy the same set of engagement functional requirements with some minor modifications and a small amount of imagination. HEW systems will have a *flyout* time that is based on the speed of light travel. Midcourse guidance/control and terminal homing will be handled by the dispensing system called aim-point control. Lethality will be a calculation based on the amount of energy over time required to damage or destroy the lethal target set. Thus, the AMD architecture functional requirements are general.

The CLS is established as a separate functional requirement within the architecture even though each functional element of the AMD architecture, with the exception of the target system, has a communication function component. Interoperability requirements will dictate a need to establish a consistent set of functional communication link and pathway standards and designs as part of the architecture. Communication standards within the architecture will flow down into the subsystem requirements of the other AMD architecture system and element requirements. The communication links will be functionally defined by operating frequency spectrum and bandwidths, word formats, data rates, and a standard frame of reference (temporally and spatially). Each communication node set will need to be identified and defined according to the functional requirements. This will ensure that interoperability requirements are satisfied.

5.3 Allocation of Functions to Systems

The functional requirements from Section 5.2 must be decomposed into functional elements or systems. A possible systems architecture flow down from the functional requirements is shown in Figure 5.1 and is the assumed architecture for the remainder of this book. The systems within the AMD systems architecture include the sensor suite system, the battle management system (BMS), and the engagement computer system (ECS). These systems will be required to satisfy the functional requirements of the CDS.

The engagement system, ISR system, CLS, and target system follow directly from the functional requirements. Functional interoperability requirements are required between geographically separated elements of the assumed AMD system architecture. Moreover, it is important to include the target system as part of the AMD architecture. There will be a set of targets within the design bounds and then there will be those outside the design bounds when the acquisition is complete. It is also an important distinction to not include the ISR suite as part of the AMD system but within the overarching AMD architecture. For example, the ISR interface should be seamless without necessarily affecting the design and independent operation of the AMD system without an ISR component. ISR is defined as an inorganic asset of the AMD architecture. Organic AMD systems include all CDS elements,

Phase A: AMD System Requirements

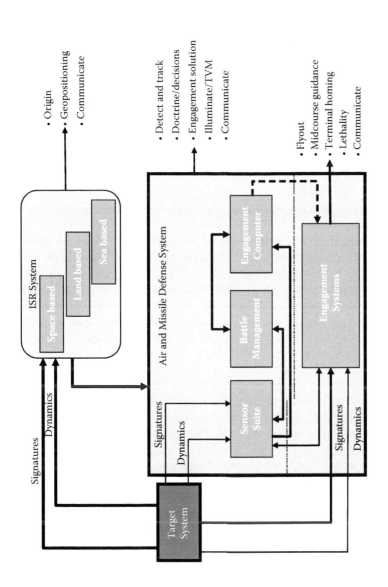

FIGURE 5.1
Assumed air and missile defense architecture.

and inorganic elements include those that may or may not be independently located from the interoperable elements that make up a complete AMD system. The engagement systems will be defined as organic elements within the AMD system even if they may operate at times independently from the remainder of the interoperable AMD system. Other definitions may be just as practical but will not be explored further.

The functional architecture identifies where systems and some elements have interoperability requirements to be defined in the next phase. An interoperability requirement exists where there are signal flow lines connecting systems or elements of systems.

The sensor suite system will be defined to primarily satisfy the detect-and-track functional requirement. However, other functional requirements may be placed on the sensor suite system as the design matures. The detect-and-track function includes a number of highly complex operations to meet the functional requirement. The sensor suite system will produce tracks on all relevant contacts while continuing to search for and detect new and potentially hostile contacts, distinguish nonthreatening contacts, produce fire-control quality data on threatening tracks, support engagements on tracks with the highest priority identification based on doctrine, and conduct resource management.

The battle management system (BMS) computes and executes the doctrine selection(s) and all decisions involved during the prosecution of the engagement. Doctrine selections and commands are then passed to the other elements as necessary for computational or execution purposes including mode initialization and weapon selection. The BMS will have responsibility to communicate selections, decisions, and computations to the other systems within the architecture. The BMS will be dependent on data transmissions from the sensor suite system, ISR, ECS, and the engagement systems both before and after launch. BMS decisions and computations affecting the engagement system will be communicated through hardwire pathways prior to weapon release and possibly through the sensor suite system or the ECS while in flight. While in flight, in this assumed architecture, the engagement system will possibly communicate information through the sensor suite system.

The ECS computes engagement solutions, sometimes referred to as a fire-control solution, leading to engageability predictions for the BMS to select firing doctrine, makes weapon selection(s), computes the predicted intercept point (PIP), and establishes a firing timeline. The ECS has communication requirements with the BMS, the sensor suite system, and the engagement systems and may be required to produce midcourse guidance commands. The dotted line connecting the ECS to the engagement system indicates that although the actual midcourse guidance or terminal homing commands are computed in the ECS, the actual communication connection is made through a data link possibly transmitted through the sensor suite system.

The ISR system produces target cueing information that may include target origin, geoposition time history, and forward prediction end goal and communicates that information to the AMD system. Cueing information will be an input to the sensor suite system to establish search volumes and improve resource management options established within the BMS and other systems. The ISR system will have to achieve latency, precision, and accuracy requirements with respect to target origin and geoposition time histories that satisfy the AMD system update requirements.

The engagement systems will require a host of weapons in an arsenal that can be selected based on performance predictions and availability. The weapons within the engagement systems can be multiple missile variants to handle short-, medium-, and long-range engagements, EW systems for shorter-range engagements, and guns for shortest-range engagements. HEW systems are not likely to be more than relatively short range and for soft targets but may be included in future arsenals. The functional strategy is to employ multiple layers of defensive weapon options whose sum effectiveness produces a high cumulative probability of kill (P_{kc}) against the required target sets. The advisable design strategy is to shoot early and to shoot often. Remember we assert the cost ratio is target-to-defended asset. The expended weapons are a relatively cheap part of the defended asset.

Communication links and pathways exist between the AMD system and the ISR system and within the AMD system as a whole. The AMD system must also receive and process potential identification friend or foe (IFF) transponder links from third-party systems as well as discriminate real from unintended targets. The BMS, the sensor suite system (SSS), the ECS, and the ES are all interconnected with either communication links, hardwired pathways, or both. Mostly, the communications will be between the BMS and the other systems and elements since the BMS is responsible for decision making and communicating the decisions where they are to be executed. This establishes another point of the architecture not necessarily obvious by the diagram. There is a requirement for an integral *man-in-the-loop* (MITL) communication source. Minimally, initialization and manual override functional requirements exist. Realistically, by the time it is necessary for an engagement to take place, it will not likely be practical for human intervention to make the countless decisions and commands necessary to have the AMD system succeed. When and where MITL is necessary and will not cause the system to fail has to be explored with detailed simulation and live exercises.

As the tactical picture evolves, updates to BMS decisions and commands will be issued. This establishes an update rate requirement that flows down to the performance requirements of each attached element. AMD system reaction time, predicted intercept point (PIP) specifications, handover error, and resource utilization are additional functional requirements of the AMD system.

The fundamental CDS performance metric is captured in the amount of time it takes to completely prosecute an engagement while accomplishing the functions necessary to meet the upper-level requirements and constraints. It is, therefore, necessary to develop a complete, but not necessarily unique, definition of CDS functionality correlated with associated systems and tied to a generalized performance timeline definition set. The notional battlespace timeline shown in Figure 5.2 assumes that there are four functional elements comprising the air and missile defense (AMD) system architecture described earlier. The CDS is composed of a sensor suite system, BMS, ECC, and an engagement system.

The SSS is responsible for detecting the target and transitioning it to track. The time it takes for this process to occur is dependent on a number of factors that will need to be itemized. They include detection range, search volume, frame time, resource allocation management, and other parameters that will be identified later.

The next step is to develop a fire-control solution (FCS). This will permit the fastest reaction time if hostile intent is determined imminently. The ECC will perform this function and pass the solution to the BMS followed by a tentative engagement order and a launcher allocation. If necessary, an identification friend or foe (IFF) challenge will be requested by the BMS. The BMS will then make a determination to prosecute the engagement or not. While these processes are taking place, the system will need to continue to operate searching, detecting, and transitioning to track potential targets. The number of simultaneous operations that can take place will be the next most important performance metric to be defined in the CDS. This is primarily specifying computer architecture, computational speed, and communication bandwidth performance once the search volume, desired total target number, and target rate specifications are determined.

The FCS is updated by the ECC, and assuming a hostile target has been determined, the BMS will produce an engagement order initiating a predicted intercept point calculation in the ECC. At this point, the BMS will need to allocate the resources necessary to fulfill the engagement order. Uplink and/or illuminator elements are possible resources that may be needed to support a missile engagement. Midcourse acceleration commands or the data necessary for the missile onboard computer to produce them will be uplinked to the missile. A semi-active terminal homing receiver will require illuminator coverage during a specified portion or for the entire terminal homing phase. Power is supplied to the missile to initialize onboard reference systems and/or navigation systems. This process typically includes the final stages of mechanical preparations that may include gyroscope stabilization and firing of squibs to uncage the seeker and/or initiate batteries and other operations such as arming of the warhead and propellant ignition. Finally, missile away includes initiating the missile boost system propelling the missile from the launch system for some specified unguided flight period where it will clear the launch system and attain a sufficient velocity to begin controlled flight.

Midcourse guidance will begin when a specified set of conditions are met whereby the missile can be controlled and guided to a predicted intercept point. One of the main criteria used for initiating midcourse guidance is reaching a dynamic pressure threshold. During this period, the CDS will

Phase A: AMD System Requirements

AMD System Function	Time	AMD Element
First detect	CDS Time Constant	Sensor Suite System
Transition to track		Sensor Suite System
Fire-control solution		Engagement Control Computer
Tentative Engagement Order (EO)		Battle Management System
Allocate launcher		Battle Management System
IFF challenge		Battle Management System
Fire-control solution		Engagement Control Computer
EO confirmed		Battle Management System
Predicted intercept point		Engagement Control Computer
FC uplink resource allocation		Battle Management
Missile initialization		Engagement Control Computer
Missile away		Battle Management System
Midcourse guidance/discrimination	Flyout Time	SSS, BM, ECC, Missile
Handover/discrimination		SSS, BM, ECC, Missile
Terminal homing/divert/discrimination		Missile
Kill evaluation		SSS, BMS, ECC, Missile

FIGURE 5.2
Notional battlespace timeline.

be required to discriminate the intended target to be destroyed from unintended tracks and intentional false targets. The sensor suite system, BMS, ECC, and missile will have computational roles to perform during this flight phase regardless of the final midcourse architecture solution chosen.

The handover doctrine will need to be finalized within the BMS as the missile approaches the PIP. Multiple terminal homing modes, preferred homing time, and approach angle decisions may be part of the handover doctrine. Discrimination decisions will need to continue within the handover doctrine and will include terminal homing sensor pointing commands. ECC will compute estimated time to go and BMS will determine handover time to go. The BMS handover decision is based on an ECC-computed P_{ssk} given measured or estimated target characteristics, geometric and environmental considerations, and engagement constraints.

Once handover has occurred, the missile terminal homing sensor provides the guidance system with target geometric and state information to close the guidance and control loop. The CDS does not contribute to guidance commands except in track-via-missile (TVM) systems architectures where CDS continues to provide guidance commands based on missile track information. Semi-active radar (SAR) missile systems rely on the CDS to point the illuminator source at the target. Both TVM and SAR systems limit the AMD system to LOS engagements only due to their continued burden on the CDS, which has a finite amount of resources, limiting the number of simultaneous engagements that can be prosecuted. Active radar and passive systems are therefore preferred assuming that all other requirements can be solved within the available technology limits.

5.4 System Performance and Interface Requirements

The key AMD system requirements can be derived from a keep-out volume requirement correlated to the MOEs found in Section 5.3. The keep-out volume can be defined as a hemispherical region that is centered on the radar array as shown in Figure 5.3.

The AMD system must be capable of kinematical engagement of targets with a given probability of single-shot engagement kill probability (P_{ssk}) within the context of the MOEs before the targets can penetrate the edge of the keep-out volume. The keep-out volume is characterized by the keep-out range (down- and cross-ground range) and the keep-out altitude vector is specified by R. The AMD system should be capable of engaging any targets that penetrate the keep-out volume boundaries.

In order to support these requirements, the radar system must be able to detect-and-track targets at ranges beyond the keep-out volume. The combination of CDS reaction time and missile flyout time to the keep-out volume boundary must be less than or equal to the time it takes the target to

Phase A: AMD System Requirements

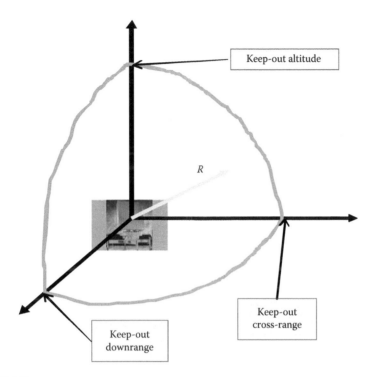

FIGURE 5.3
Illustration of the keep-out volume concept.

reach the defended volume boundary once placed in firm track. Transition of detected target to firm track usually requires some statistical number of correlated target detections on successive confirmation dwells. One can quickly see the importance of balancing missile flyout range and speed with the radar detection range to achieve a robust AMD system design that supports the keep-out volume requirement. The inbound target defense penetration techniques described in Section 4.4.3, including most importantly speed, will be a significant driver on required radar and missile performance. Early target detection and location will be a fundamental performance requirement.

The ISR system will communicate to the BMS and ultimately cue the sensor suite system where the search radar will be assigned a search sector and other performance parameters. ISR performance requirements such as track accuracy and latency will be established to support the development of a successful system.

5.4.1 Central Defense System Performance Requirements

The primary purpose of the CDS is to deliver weapons to engage, intercept, and destroy targets. Destroying the intended target requires achieving an acceptable miss distance as dictated by the kill strategy, kill mechanism, and target vulnerability. The CDS must also conduct the assigned tasks shown in Figure 5.2 and do so in a timely manner so as to maximize the battlespace thereby permitting a maximum number and variant of weapons to be used to suppress the target. This problem is known as reaction time trade space. The reaction time trade space can include reducing the number of processes in the battlespace timeline, reducing the amount of time it takes to accomplish specific processes, or a combination of both.

The CDS performance requirements will be quantified by accomplishing these tasks within a given timeline while achieving specified accuracy objectives. Functional requirements will be quantified by tasks with associated accuracy specifications to establish performance requirements and are described in the following paragraphs.

Achieving an acceptable miss distance burdens the CDS with (1) accurately tracking the target(s) and launched missile(s), (2) accurately locating itself in the inertial reference coordinate frame, (3) accurately aligning the CDS navigation and missile flight coordinate frames and the missile inertial reference unit (IRU) prior to launch, (4) accurately computing a predicted intercept point (PIP), (5) providing the missile accurate midcourse guidance commands, and (6) accurately pointing the missile's homing sensor–sensitive axis toward the target prior to handover.

IRU alignment greatly influences missile midcourse performance. Selecting alignment techniques and IRU components is a requirement of trade-space study that dominates position, velocity, and altitude errors affecting seeker pointing angle accuracy when the missile sensor (seeker) attempts target acquisition. These accumulated errors then translate into expanded field-of-view and acquisition range requirements. In turn, the alignment techniques that produce higher accuracy drive up both cost and, more importantly, time to alignment completion increasing the CDS time constant.

The missile IRU is typically aligned with the CDS position vector in the inertial space reference frame. The accuracy of the CDS position knowledge introduces additional error. Modern platforms typically use some type of satellite-aided inertial navigation like the global positioning system (GPS) for alignment. The missile midcourse error is also affected by radar track error. The CDS sensor suite may be required to produce continually updated skin tracks of the missile in flight as a source for missile position, velocity, and altitude. These measurements along with the missile onboard guidance computer–processed IRU measurements are combined using filtering techniques to establish the best estimate of missile in-flight states and time to go to predicted intercept point. Transponder techniques are normally used to transmit missile onboard measurements to CDS for processing. The error source summary is itemized as follows:

- Missile-in-flight guidance (IRU measurements and state computations)
- Platform postlaunch drift (probably negligible)

Phase A: AMD System Requirements

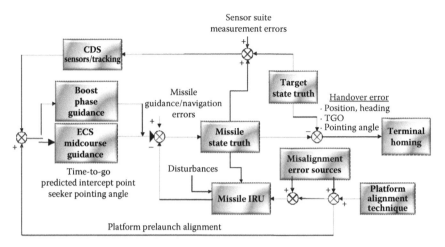

FIGURE 5.4
Midcourse guidance and error contributions to handover.

- CDS sensor suite measurement
- Missile prelaunch misalignment
- Platform prelaunch navigation

The midcourse navigation process is depicted in Figure 5.4. Boost phase guidance precedes midcourse by some predetermined time based on the missile boost design and trajectory requirements. Figure 5.4 will be explained in detail in the following paragraphs and sections [8–23,25,27].

5.4.1.1 Midcourse Guidance Reference Systems

Coordinate systems are defined here to conform to a standard throughout the book and to adhere to a common standard within the aeronautics community such as the AIAA standards [28]. Four right-handed, Cartesian coordinate reference systems pertaining to the CDS for missile midcourse guidance are defined. These are the earth-centered earth-fixed (ECEF) Cartesian reference system (X_e, Y_e, Z_e) axes, the midcourse guidance or navigation reference system (N, E, D), the missile body reference system (x_b, y_b, z_b) axes, and the seeker reference system (a_s, b_s, c_s) axes as depicted in Figure 5.5. The midcourse guidance reference system is sometimes referred to as the launch-centered inertial Cartesian (LCIC) reference system axes. LCIC will be used in this book.

It is assumed that the target engagement or flyout phase will be composed of four segments defined here as prelaunch, boost, midcourse, and terminal. These segments are defined in Figure 5.6 and the following paragraph.

The prelaunch segment definition is given within the CDS time constant shown in Figure 5.2 and discussed in the previous section. The boost,

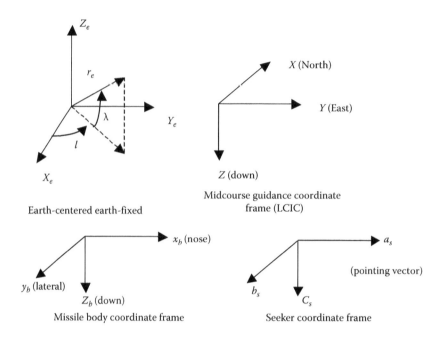

FIGURE 5.5
CDS requirement coordinate frames.

midcourse, and terminal segments are defined within the context of the flyout time also shown in Figure 5.2. Two transition stages are identified in Figure 5.5. The first transition stage is booster engine cutoff (BECO) marking the end of the boost phase guidance and the burnout of the initial booster and the beginning of the midcourse guidance/navigation phase. The second transition phase is handover. Handover marks the end of midcourse guidance/navigation and the beginning of the terminal homing phase.

CDS performance requirements at handover, which precedes the final segment, terminal homing, and engagement segment, are defined next. In the engagement concept of operation, the missile transitions to an onboard homing sensor to execute the terminal segment and to achieve the designed kill criteria including kill strategy and miss distance performance specifications. Engagement system performance requirements are driven by the terminal homing phase. Handover error will be a key contributor to miss distance performance, and itemizing the error contributions will be the primary focus of the CDS performance requirements. It is first necessary to fully characterize the geometric relationships of the engagement. Figure 5.7 provides detailed in-flight midcourse guidance coordinate system relationships and definitions.

The geometric relationships shown in Figures 5.5 through 5.7 are defined through three direction cosine matrices transforming vectors from the ECEF

Phase A: AMD System Requirements

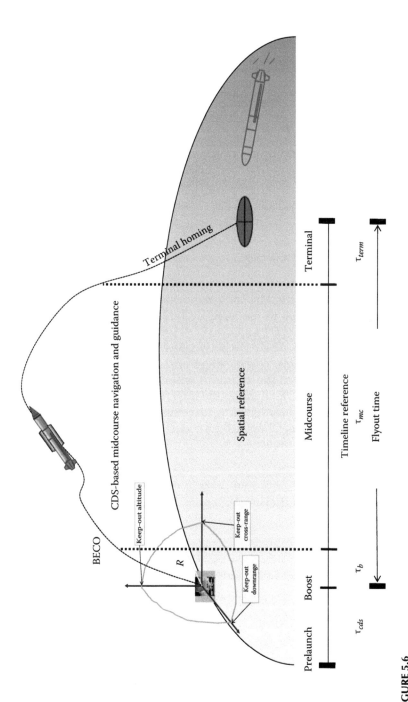

FIGURE 5.6
Engagement segment definition.

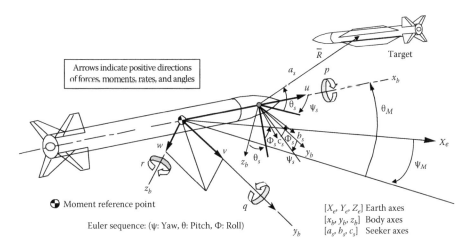

FIGURE 5.7
Midcourse guidance terms and definitions.

system to the body axis, from the body axis system to the seeker axis system, and from the LCIC system to the ECEF system. These transformations are denoted by $A = [T^b_{ECEF}]$, $B = [T^s_b]$, and $C = [T^{ECEF}_n]$, respectively, and provided in Equations 5.1 through 5.6, where b = body reference, n = LCIC reference, and s = seeker reference systems. When using transformations to move from one reference frame to another, the subscript will denote the original frame of reference and the superscript denotes the final frame of reference. For example, T^b_n refers to a matrix accomplishing transformation from the LCIC frame to the missile body frame. Chapter 8 provides the mathematical details for deriving these transformations:

$$[A] = \begin{bmatrix} a_{11} & a_{12} & a_{13} \\ a_{21} & a_{22} & a_{23} \\ a_{31} & a_{31} & a_{33} \end{bmatrix} \tag{5.1}$$

$a_{11} = \cos\Theta_M \cos\Psi_M$

$a_{12} = \cos\Theta_M \sin\Psi_M$

$a_{13} = -\sin\Theta_M$

$a_{21} = \sin\Phi_M \sin\Theta_M \cos\Psi_M - \cos\Phi_M \sin\Psi_M$

$a_{22} = \sin\Phi_M \sin\Theta_M \sin\Psi_M + \cos\Phi_M \cos\Psi_M$

$a_{23} = \sin\Phi_M \cos\Theta_M$

$a_{31} = \cos\Phi_M \sin\Theta_M \cos\Psi_M + \sin\Phi_M \sin\Psi_M$

$a_{32} = \cos\Phi_M \sin\Theta_M \sin\Psi_M - \sin\Phi_M \cos\Psi_M$

$a_{33} = \cos\Phi_M \cos\Theta_M$

(5.2)

$$[B] = \begin{bmatrix} b_{11} & b_{12} & b_{13} \\ b_{21} & b_{22} & b_{23} \\ b_{31} & b_{31} & b_{33} \end{bmatrix} \quad (5.3)$$

$$\begin{aligned} b_{11} &= \cos\Theta_s \cos\Psi_s & b_{21} &= -\sin\Psi_s & b_{31} &= \sin\Theta_s \cos\Psi_s \\ b_{12} &= \cos\Theta_s \sin\Psi_s & b_{22} &= \cos\Psi_s & b_{32} &= \sin\Phi_s \sin\Psi_s \\ b_{13} &= -\sin\Theta_s & b_{23} &= 0 & b_{33} &= \cos\Theta_s \end{aligned} \quad (5.4)$$

$$[C] = \begin{bmatrix} c_{11} & c_{12} & c_{13} \\ c_{21} & c_{22} & c_{23} \\ c_{31} & c_{31} & c_{33} \end{bmatrix} \quad (5.5)$$

$$\begin{aligned} c_{11} &= \cos\ell & c_{21} &= 0 & c_{31} &= -\sin\ell \\ c_{12} &= -\sin\lambda\sin\ell & c_{22} &= \cos\lambda & c_{32} &= -\sin\lambda\cos\ell \\ c_{13} &= \cos\lambda\sin\ell & c_{13} &= \sin\lambda & c_{33} &= \cos\lambda\cos\ell \end{aligned} \quad (5.6)$$

5.4.1.2 Handover

Handover performance parameters are defined here as the spatial, temporal, and kinematic set of engagement terms that are required by the weapon and provided by the CDS to acquire and track the target and sufficiently reduce the miss distance to destroy the target. Essentially, the CDS is responsible for midcourse guidance commands that place the missile at a particular spatial coordinate and time (time to go [TGO]), with a specified velocity vector (orientation and magnitude) while aiming the sensor-sensitive axis at the target. Velocity vector orientation will be referred to here as heading. Midcourse guidance sends the missile toward the predicted intercept point (PIP) but hands over the engagement to the missile at a computed time to go prior to reaching the PIP that computationally maximizes the probability that the missile will achieve a minimum miss distance from the target at the closest point of approach thus maximizing the single-shot probability of kill (P_{ssk}). Handover geometry definitions are shown in Figure 5.8 and are described in the following paragraphs.

The line-of-sight (LOS) vector has an argument, λ_s, with the magnitude being the distance between the missile and the target. The arguments HE (heading error) and η (seeker pointing angle error) are specified as midcourse guidance handover error requirements. HE and η specifications are functions of other engagement variables that include target range, altitude,

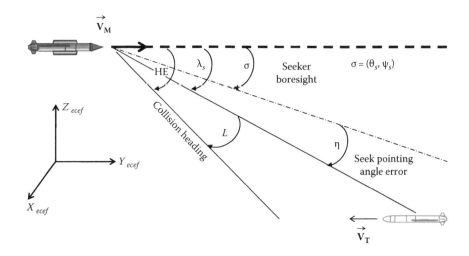

FIGURE 5.8
Handover geometry.

velocity, range to go, time to go, and missile velocity at handover. Velocity magnitude (speed) is specified as the optimally largest achievable value for the engagement. A handover error requirements trade study is a required part of the CDS engagement system trade space necessary to balance overall ADS performance requirements. Once completed, the trade space results from the performance system requirements will flow down to the subsystem performance requirements.

Heading error (HE) can be defined mathematically as the difference between the true V_m direction and the required V_m direction necessary to achieve a zero line-of-sight (LOS) rate (proportional navigation principle). Lead angle ($\mathcal{L}$) is defined as the angle between the target velocity vector and the angle required to perfectly lead the target to obtain a collision. Refer to the classical collision triangle in Figure 5.9 for the analysis that follows [8].

The initial missile and target positions in the general X–Y axis coordinate frame are established next. The missile is initially located at the origin and is shown as (M_{x0}, 0), and the initial target position is given as (T_{x0}, 0). The **LOS** between the target and the missile is **T** − **M**. The intercept point **I** is collocated where **T** − **M** = 0 to satisfy the conditions for the intercept. When the missile and target are considered to be flying at constant velocities, then the condition $T_2 - M_2 < T_1 - M_1 < T_0 - M_0$ is satisfied and successive similar triangles are formed producing successive nonrotating **LOS** angles in time. Therefore, this proof shows that the two conditions sufficient for intercept are a nonrotating **LOS** and $d|T(t) - M(t)|/dt < 0$.

The collision triangle in Figure 5.9 can be used to compute that the $d\vec{M}/dt$ and $d\vec{T}/dt$ terms that represent missile and target velocity vectors,

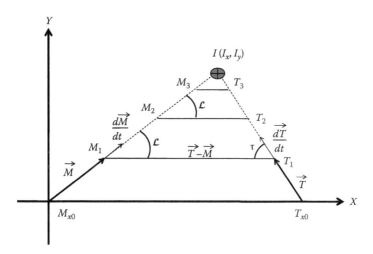

FIGURE 5.9
Classical intercept collision triangle.

respectively, have components perpendicular to the **LOS** and **(T − M)** that are equal. Therefore, $\mathcal{L}$ is the required lead angle.

Again, using Figure 5.9, the mathematical definition for $\mathcal{L}$ is formulated:

$$\mathbf{V_m} \sin(\mathcal{L}) = \mathbf{V_t} \sin(\tau) \tag{5.7}$$

Solving Equation 5.7 for $\mathcal{L}$ provides the expression for the required lead angle:

$$\mathcal{L} = \sin^{-1}(\mathbf{V_t}/\mathbf{V_m} \sin(\tau)) \tag{5.8}$$

Figure 5.8 and Equation 5.8 allow us to establish the CDS performance requirements that are given in Equations 5.9 and 5.10:

$$HE = \lambda_s + \mathcal{L} \tag{5.9}$$

$$\lambda_s = \sigma + \eta \tag{5.10}$$

The handover requirements' trade space includes seeker pointing angle error (η), seeker field of view (FOV), and acquisition range. The triad requirement to be specified is [η, FOV, R_{acq}]. Minimizing η within the seeker design constraints [FOV, R_{acq}] will establish the probability of target acquisition that will be specified as a seeker design requirement. The seeker requirement goal is proposed to minimize the R_{acq} requirement. This approach relieves seeker transmitter power and/or operating frequency requirements while still providing countermeasure resistance and smaller achievable miss distances. Subsequently, this requirement philosophy will place the bulk of the handover requirement on the CDS performance requirements and demand a highly accurate platform navigation and midcourse guidance segment. Missile flow down performance requirements will include specifying minimum *Tgo* and kinetic energy (velocity) requirements to ensure that maneuverability requirements are met in terminal homing.

5.4.1.3 Seeker Pointing Angle Error

The seeker pointing angle error, η, specification should be developed first. The techniques and results of [8–11] are used here to quantify (η). How to compute the seeker or onboard missile sensor pointing angle error is derived and fully explained with examples in [8,9]. The results here follow the cited references with the exception of coordinate frame definitions and derivations as applicable.

The two primary contributors to seeker pointing angle error, η, are missile midcourse guidance/navigation and target measurement errors. To derive η, additional terms need to be defined. The relative range vector between the target and the missile will be **R**, where **R** = **T** − **M**. Then the seeker pointing unit vector in body frame coordinates is

$$\vec{P}^b = \frac{\vec{R}^b}{|R^b|} \quad (5.11)$$

A close-up view of Figure 5.7 is presented in Figure 5.10 to more clearly establish the seeker axis-to-missile body axis relationship.

The orthogonal components of x_s, y_s, and z_s will be the axis about which the critical seeker pointing angle error component requirements are specified. Error components (y_s, z_s) are determined, and then the angular error associated with this axis is the seeker pointing angle error η = ($η_y$, $η_z$). During midcourse guidance, the body coordinate pointing vector, P^b, is found by computing the relationships in Equations 5.12 and 5.13:

$$D = [A][C] \quad (5.12)$$

$$P^b = [D][(\mathbf{T}^n - \mathbf{M}^n)/|[D](\mathbf{T}^n - \mathbf{M}^n)|] \quad (5.13)$$

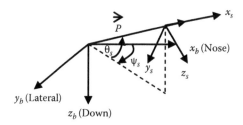

FIGURE 5.10
Missile body and seeker coordinate frame relationship.

Phase A: AMD System Requirements

The missile guidance computer receives the term $(\mathbf{T}^n - \mathbf{M}^n)$ computed by the ECS at a design update rate and then completes the Equations 5.12 and 5.13 computation. This process occurs just prior to handover to begin the seeker acquisition phase. The seeker pointing angle error vector, $\boldsymbol{\eta}$, resulting from this process is completely derived with examples in [8,10]. The derivation is beyond the scope of this book, but the reader is referred to those references for a thorough understanding of the process to compute $\boldsymbol{\eta}$, the resulting seeker pointing angle error components (η_y, η_z).

A diagram defining the various required performance parameters is introduced in Figure 5.11 from the cited reference. Note that the seeker pointing angle error, η^s, is measured relative to the missile-to-target LOS, and therefore, the x_s axis component of η^s is zero.

The following definitions apply:

$$R^b \stackrel{\Delta}{=} \begin{bmatrix} R_x^b \\ R_y^b \\ R_z^b \end{bmatrix}; \quad \delta R_n^b \stackrel{\Delta}{=} \begin{bmatrix} \delta R_{nx}^b \\ \delta R_{ny}^b \\ \delta R_{nz}^b \end{bmatrix}; \quad \delta p_b^s \stackrel{\Delta}{=} \begin{bmatrix} \delta p_{bx}^s \\ \delta p_{by}^s \\ \delta p_{bz}^s \end{bmatrix}; \quad p^b \stackrel{\Delta}{=} \begin{bmatrix} c\theta_s c\psi_s \\ c\theta_s s\psi_s \\ -s\theta_s \end{bmatrix}; \quad P^s \stackrel{\Delta}{=} \begin{bmatrix} T_b^s p^b \end{bmatrix} = \begin{bmatrix} 1 \\ 0 \\ 0 \end{bmatrix}$$

$$\begin{bmatrix} \delta R_{nx}^b \\ \delta R_{ny}^b \\ \delta R_{nz}^b \end{bmatrix} \stackrel{\Delta}{=} \begin{bmatrix} \delta T_{nx}^b - \delta M_{nx}^b \\ \delta T_{ny}^b - \delta M_{ny}^b \\ \delta T_{nz}^b - \delta M_{nz}^b \end{bmatrix}; \quad \eta^s \stackrel{\Delta}{=} \begin{bmatrix} \eta_x^s \\ \eta_y^s \\ \eta_z^s \end{bmatrix} = \begin{bmatrix} 0 \\ -\delta p_{bz}^s \\ \delta p_{by}^s \end{bmatrix}; \quad \phi^n \stackrel{\Delta}{=} \begin{bmatrix} \phi_x^n \\ \phi_y^n \\ \phi_z^n \end{bmatrix}$$

where the components of η^s are small angular rotations defined as positive counterclockwise shown in Figure 5.11 and ϕ, the error vector tilt angle, relates the error-free midcourse guidance frame, n, to the computed CDS platform attitude reference coordinate system.

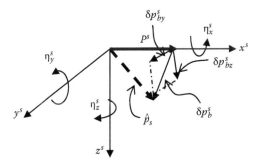

FIGURE 5.11
Pointing angle error definitions and terms. (Derived from Kouba, J.T. and Bose, S.C., *IEEE Transactions on Aerospace and Electronic Systems*, AES-16(3), 313, May 1980 [10].)

To complete the seeker pointing angle error results, it is necessary to define the skew symmetric matrices $[H^s]$ and $[F^n]$:

$$H^s \stackrel{\Delta}{=} \begin{bmatrix} 0 & -\eta_z^s & \eta_y^s \\ \eta_z^s & 0 & -\eta_x^s \\ -\eta_y^s & \eta_x^s & 0 \end{bmatrix}; \quad F^n \stackrel{\Delta}{=} \begin{bmatrix} 0 & -\phi_z^n & -\phi_x^n \\ \phi_z^n & 0 & -\phi_x^n \\ -\phi_y^n & \phi_x^n & 0 \end{bmatrix}$$

The seeker pointing angle error equations [9] can now be written as shown in Equations 5.14 and 5.15:

$$\eta_y^s = [(\delta M_{nx}^b - \delta T_{nx}^b)/|R|]\sin\theta_s \cos\psi_s + [(\delta M_{ny}^b - \delta T_{ny}^b)/|R|]\sin\theta_s \sin\psi_s$$
$$+ [(\delta M_{nz}^b - \delta T_{nz}^b)/|R|]\cos\theta_s - \phi_x^b \sin\psi_s + \phi_y^b \cos\psi_s \quad (5.14)$$

$$\eta_z^s = [(\delta T_{ny}^b - \delta M_{ny}^b)/|R|]\cos\psi_s + [(\delta T_{nx}^b - \delta M_{nx}^b)/|R|]\sin\psi_s$$
$$+ \phi_x^b \sin\theta_s \cos\psi_s + \phi_y^b \sin\theta_s \sin\psi_s + \phi_z^b \cos\theta_s \quad (5.15)$$

Kouba and Bose [10] offer a seeker pointing angle error computational process shown in Figure 5.12.

5.4.1.4 Midcourse Guidance

The function of midcourse guidance is to minimize energy loss prior to terminal homing to ensure that the energy demands of terminal homing are met despite heading, seeker pointing errors, and demanding target kinematics. In order to fully understand the requirements associated with midcourse guidance, additional terms must be understood. From Figure 5.8, the geometry is modified slightly to obtain the following variable definitions:

- R_{MT}: Range between the target and the missile along LOS
- R_{Tpip}: Range to the PIP from the current target position
- R_T: Current target position (LCIC)
- R_{Mpip}: Missile range to go to the PIP
- R_{pip}: PIP range vector in the LCIC frame
- R_M: Missile range vector in the LCIC
- Tgo: Missile time to go to the PIP
- V_M: Current missile velocity magnitude
- V_{ML}: Missile velocity along the PIP LOS
- V_{TM}: Target to missile velocity along the LOS

Phase A: AMD System Requirements

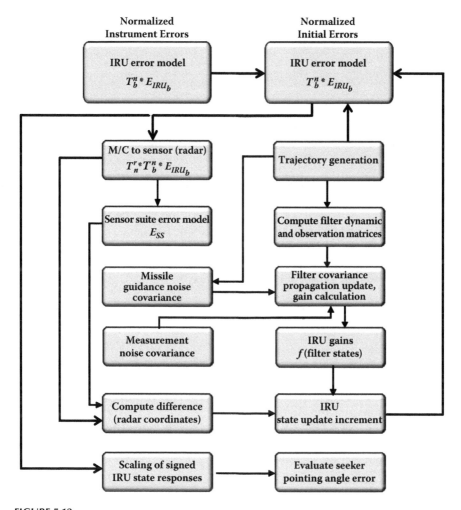

FIGURE 5.12
Seeker pointing angle error computational process. (From Kouba, J.T. and Bose, S.C., *IEEE Transactions on Aerospace and Electronic Systems*, AES-16(3), 313, May 1980 [10].)

- V_{MF}: Missile velocity vector at the PIP
- θ_T: Angle between R_{MT} and V_T
- λ: Target to missile LOS angle
- γ_M: Inertial angle to collision course with the PIP
- θ_M: $\mathcal{L}$ (angular deviation between the LOS and collision point)
- $\delta =$ HE for V_M (angular deviation from collision course)
- $\mu =$ Velocity angle error of the present and final missile velocity vector

Then, the following relationships hold to define the PIP:

$$V_{ML} = V_M R_{Mpip} \cos(\delta) \tag{5.16}$$

$$\theta_M = \gamma_M - \lambda \tag{5.17}$$

$$\theta_M = \sin^{-1}(V_T \sin(\theta_T)/V_{ML}) \tag{5.18}$$

$$Tgo = R_{Mpip}/V_{ML} \tag{5.19}$$

$$R_{Tpip} = R_T + (V_T)Tgo \tag{5.20}$$

The midcourse guidance problem is a two-point boundary value problem associated with placing the missile at the PIP using energy management control [12]. This is known as explicit guidance because the current and desired boundary conditions are specified. The kappa midcourse guidance produces a suboptimal minimized energy solution where the acceleration command is given by Equation 5.2.1 [12,25]. Lin [12] provides a complete derivation and discussion of kappa guidance:

$$a_c = \frac{K_1}{Tgo^2}\left[\vec{R}_{PIP} - \vec{R}_M - \vec{V}_M Tgo\right] + \frac{K_2}{Tgo}\left[\vec{V}_{MF} - \vec{V}_M\right] \tag{5.21}$$

Alternatively, Serakos and Lin [13] provide a closed-form solution of a linearized kappa guidance law that shows relatively close performance to the original kappa solution that could also be considered in design to satisfy midcourse guidance requirements.

The K terms in Equation 5.21 are time-varying gains that are selected to maximize V_{MF}. Once solved, the kappa, optimum normal acceleration command can be rewritten as follows:

$$a_c = -\frac{K_1}{R_{Mpip}}\left[\vec{V}_M^2 \sin(\delta)\right] + \frac{K_2}{R_{Mpip}}\left[\vec{V}_M^2 \sin(\mu)\cos(\delta)\right] \tag{5.22}$$

The kappa acceleration command of Equation 5.22 has two components. The first term on the right side of Equation 5.22 is a proportional navigation component, and the second term is a trajectory-shaping component. Trajectory shaping is typically only performed in the vertical plane. The gains K_1 and K_2 are written in Equations 5.23 and 5.24:

$$K_1 = \frac{\omega^2 R^2_{Mpip}(\cosh(\omega R_{Mpip}) - 1)}{\omega R_{Mpip} \sinh(\omega R_{Mpip}) - 2(\cosh(\omega R_{Mpip}) - 1)} \tag{5.23}$$

$$K_2 = \frac{\omega^2 R^2_{Mpip} - \omega R_{Mpip}(\sinh(\omega R_{Mpip}))}{\omega R_{Mpip} \sinh(\omega R_{Mpip}) - 2(\cosh(\omega R_{Mpip}) - 1)} \tag{5.24}$$

Phase A: AMD System Requirements

The variable "ω" is known as the trajectory-shaping coefficient and is a first-order function of the aerodynamic and propulsive forces in that plane when attempting to minimize energy loss for a missile flight. Using optimal control techniques [12], the derived term "ω" is given in Equation 5.25:

$$\omega^2 = \frac{D_0 L_\alpha \left(\dfrac{Thrust}{L_\alpha} + 1\right)^2}{mass^2 V_M^4 \left(\dfrac{Thrust}{L_\alpha} + 2\eta C_{L\alpha}\right)} \quad (5.25)$$

where
- D_0 is the zero-lift ($\alpha = 0$, for supersonic axisymmetric vehicles) dimensional drag force
- L_α is the lift-curve slope with angle of attack (α)
- η is an aerodynamic efficiency factor that can be determined empirically
- $C_{L\alpha}$ is the dimensionless lift-curve slope
- Thrust is the time-dependent missile thrust
- mass is the time-dependent missile mass

The midcourse guidance algorithm must involve the kill strategy through defining the terminal flight path angle constraints necessary to transition to terminal homing. The kill strategy will be influenced by whether the mission is air or ballistic targets and by whether the engagement is low altitude, endoatmospheric, or exoatmospheric. Other factors to consider are target vulnerability, fuzing considerations, and engagement geometry. These factors will be part of the requirement's trade study process. Another contribution to engagement and kill strategy will be the estimated time-to-go accuracy. There are many approaches proposed in the literature [14,15] to compute this parameter, but it will need to follow from a requirement study that includes specific target features, capabilities, and performance limitations. It is assumed that the designer understands the full spectrum of his or her own missile, and this component of the problem should be academic.

A missile simulation with a variation of kappa midcourse guidance was constructed, and Figure 5.13 presents an example of some flyouts demonstrating vertical plane trajectory-shaping strategies for a variety of engagement situations.

The strategies employed in the examples shown in Figure 5.13 are simple compared to the range of possibilities that can be explored to satisfy requirements. This strategy shows that in engagements below 10,000 m altitude, a direct 75° angle approach to the target is chosen that may be appropriate for a cruise missile defense strategy. Engagements at 10,000 m altitude and above employ an inverse trajectory approach to the target of 45° that may be appropriate for a terminal ballistic missile defense strategy. Figure 5.14 shows the time-of-flight variations associated with the trajectory-shaping variations.

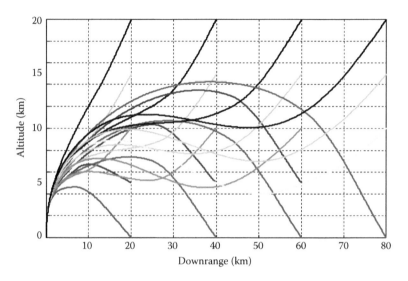

FIGURE 5.13
Trajectory shaping using kappa midcourse guidance variations.

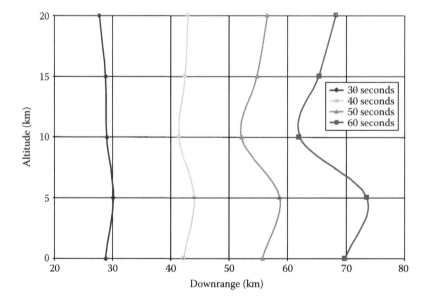

FIGURE 5.14
Kappa trajectory-shaping time-of-flight variations.

From the results shown in Figures 5.13 and 5.14, flyout tables can be constructed and used when developing engagement subsystem flow down requirements.

5.5 CDS: Sensor Suite System Performance Requirements

The following discussion will focus on developing radar requirements for detecting targets at a given range and developing and understanding the sensor suite system–level performance trade-offs for achieving a given detection range. Figure 5.15 shows the process for allocating missile defense system requirements to the radar system given a keep-out volume requirement for engaging targets in operational environments.

The key radar design parameters are detection range, range accuracy, angle accuracy, and the time it takes for searching out the surveillance volume. The allocation begins with CDS specifications and combined

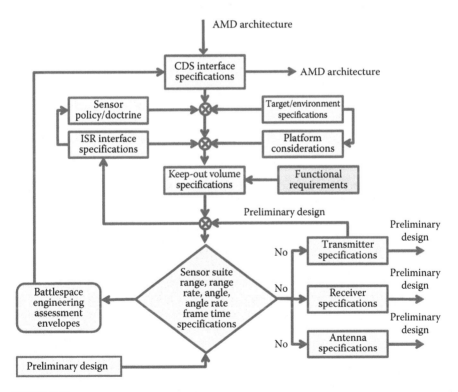

FIGURE 5.15
Process for the allocation of requirements to the radar system.

with sensor suite doctrine specifications, ISR specifications, target and environment specifications, and platform specifications. The keep-out volume is specified from the functional requirements when these components are combined.

The detection range, which is related to the firm track range, must support the keep-out volume requirement. The radar detection range can be readily determined from the radar range equation, which is a function of key radar performance parameters. Key radar performance parameters can be selected in order to achieve a detection range that supports the keep-out volume when combined with CDS time constant and the engagement system boundaries (to be discussed in Section 5.4). Moreover, the range and angle accuracy must support missile seeker handover requirements. The radar needs to guide the missile to a point in space such that when the missile goes into terminal guidance mode, the target is contained in the missile seeker field of view.

The radar architecture can be based on a rotating antenna or phased array. A rotating antenna radar is typically a 2D radar with a wide elevation beam that determines target range and azimuth. A phased array radar is a 3D radar that incorporates an electronically scanned beam to determine the target location in range, azimuth, and elevation. A tracking filter processes successive target position updates to improve the accuracy of the location estimates and estimates the target velocity and acceleration components. A rotating phased array radar scans the pencil beam in azimuth mechanically while the beam is electronically scanned in elevation.

A phased array radar antenna can be passive or active. Passive phased arrays have a separate transmitter that usually incorporates microwave tubes such as traveling wave tubes (TWTs) or crossed-field amplifiers (CFAs). An example is the SPY-1 Radar, which has a multistage transmitter that incorporates both TWTs and CFAs. Active phased arrays incorporate transmit/receive (T/R) modules that are integral to the antenna and located behind each antenna element.

A notional architecture for the passive phased array of three elements is shown in Figure 5.16. This architecture is scalable to phased arrays with thousands of radiating elements. A waveform generator (WFG) generates and provides a low-power radar pulse to the centralized transmitter that is typically a microwave tube–based transmitter. The transmitter amplifies the radar pulse to high power. Next, the high-power radar signal is divided and distributed to the individual radiating elements by the transmit beamformer. The transmit beamformer typically provides equal power to each element by using a waveguide power splitting network. The circulator at each element provides isolation between transmit and receive paths to help protect the receiver during high-power pulse transmission. Next, the radar pulse passes through a phase shifter. The phase shifters are used to apply a linear phase gradient across the array face, steering the beam in a desired direction for the transmit frequency. Phase shifters are typically narrowband and need to be reset if the radar frequency is changed significantly to keep the antenna

Phase A: AMD System Requirements

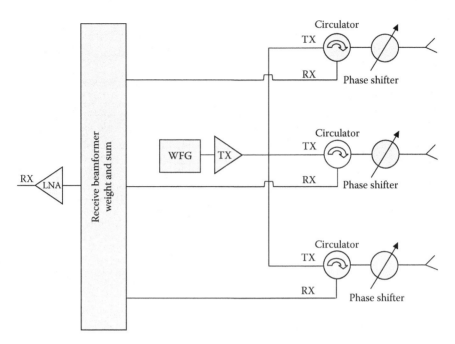

FIGURE 5.16
Notional passive phased array architecture.

beam steered in the same direction. The phase shifters are also reset to steer the beam to a new beam position.

On receive, the phase shifter settings are maintained to keep the receive beam aligned with the transmit beam. The receive signal is directed by the circulator to the receive beamforming network, which is a low-power microwave network. The receive beamformer typically applies a low sidelobe weighting where the element locations moving toward the edge of the array are more strongly weighted than those in the center of the array. The receive beamformer for a passive low sidelobe phased array typically has high losses due to the implementation of the low sidelobe weighting. This large loss ahead of the low-noise amplifier (LNA) limits the radar noise figure and sensitivity. The LNA output is first downconverted to an intermediate frequency or baseband and then converted to digital signal by the A-to-D converter for subsequent digital signal processing.

The architecture for a notional three-element active phased array is shown in Figure 5.17. The transmit and receive functions are distributed and contained in transmit/receive (T/R) modules that are located in close proximity to the radiating elements. The WFG output is now divided and distributed to each T/R module. Both transmit and receive losses are significantly reduced compared to the passive phased array architecture. The receive weighting for low sidelobes now occurs after the LNA supporting a lower system noise

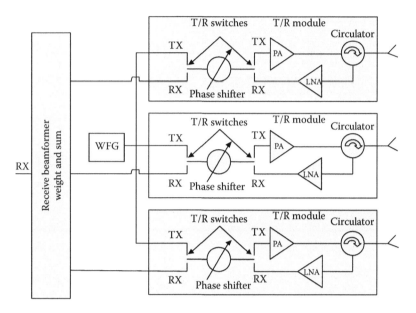

FIGURE 5.17
Notional active phased array architecture.

figure and increased radar sensitivity. Because of these architectural differences, an active phased array typically supports a 10 dB improvement in radar sensitivity compared to a passive array.

Ultimately, radar performance will be limited by prime power, cooling, and weight constraints associated with the platform (e.g., shipborne, airborne, or mobile land-borne). The radar efficiency, which is defined as the ratio of radiated power to required prime power, affects both prime power and cooling requirements. Weight must be minimized for airborne and mobile land-borne radar systems. Weight is less of an issue for shipborne radar systems; however, the weight will impact how the radar antenna can be mounted on a ship. It is desirable for a shipborne antenna to be mounted as high as possible to maximize the radar horizon for low-flying targets.

The radar range equation for a clear environment from Blake [26] is

$$R_{max} = [(P_t G_t G_r \sigma \lambda^2 F_t^2 F_r^2)/(4\pi)^3 (S/N)_{min} k T_s B_n L]^{1/4} \qquad (5.26)$$

The maximum range is a statistical quantity that is coupled to the minimum required signal-to-noise (S/N) ratio. Typically, a detection range probability of 90% is used. To maximize the radar detection range, it is desired to maximize terms in the numerator while minimizing terms in the denominator. The target radar cross section, σ, is based on the required targets to be engaged.

One of the first steps is to select a radar frequency band. The target radar cross section (RCS) is a function of aspect angle, frequency, and polarization. Therefore, the target RCS is somewhat variable and statistical in nature. The pattern propagation factor is a transmit loss (F_t^2) and a receive loss (F_r^2) that results from multipath interference. These two quantities are generally less than unity although under certain conditions, multipath can be constructive. The level of interference depends on the terrain, propagation conditions, radar frequency, radar height, and target altitude. The pattern propagation factor is generally worse at lower radar frequencies, lower radar heights, and lower target altitudes under standard propagation conditions. For shipborne applications, X-band radars are generally preferred for the detection of low-altitude targets. X-band extends from 8 to 12 GHz. X-band radars are also generally smaller and lighter than S-band radars, so they can be placed higher on the ship's structure. S-band extends from 2 to 4 GHz. Pattern propagation factors for typical S- and X-band antenna heights for shipborne installations are illustrated in Figure 5.18 as a function of target height and range [27].

Once a frequency band is selected, the antenna size (or gain) and transmitter power can be traded to achieve the required detection range to support the keep-out volume requirement. The radar detection range required to support the keep-out volume requirement is also dependent on the minimum target RCS, maximum inbound target velocity, and missile speed. For example, a notional S-band (~3 GHz) requirement is used to detect a −20 dBsm target at a range of 50 km in the presence of a 20 dB two-way pattern propagation factor loss. The radar pulse width, τ, is 50 µs and the losses, L, are 6 dB. In addition, the minimum required signal-to-noise ratio is 13 dB and the system noise temperature, T_s, is 30 dB, which is a typical value for passive phased array radars. The fixed parameters for a notional passive phased array radar are summarized in Table 5.2. This requires the combination of transmit power, transmit gain, and receive gain ($P_t G_t G_r$) to be 144.4 dB.

Allocating this requirement equally among the transmit power and transmit and receive gain results in a transmitter peak power requirement of 65 kW and an antenna size of 51.8 m² or a diameter of 8.1 m. Clearly, an antenna diameter of 8.1 m is not practical. However, a transmitter with a peak power level of 1 MW is practical using crossed-field amplifier (CFA) microwave tube technology. This results in an antenna size of 13.2 m² or a diameter of 4.1 m, which may be possible for a shipborne system. A 13.2 m² antenna would contain approximately 4600 elements in a triangular lattice. The use of a triangular lattice versus a rectangular lattice reduces the number of elements required by approximately 13% [29]. The required peak power per element is 217 W.

Several other techniques can be used to increase radar sensitivity. The pulse width, τ, can be increased or multiple pulses can be integrated. Both approaches have radar performance trade-offs. For example, pulse widths as

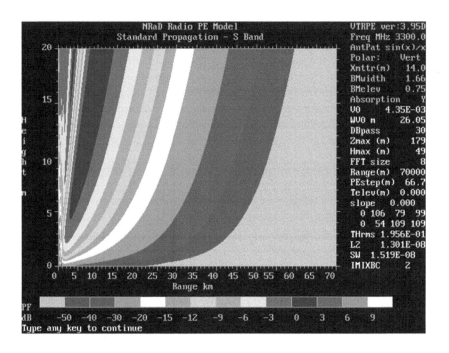

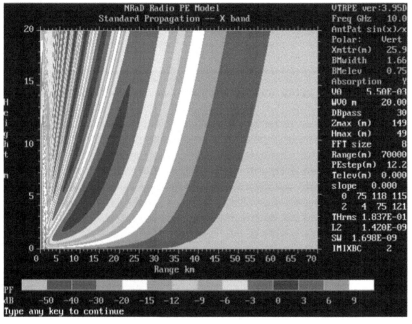

FIGURE 5.18
Pattern propagation factor comparison of an S-band (antenna height 14 m) and X-band (antenna height 26 m) in dB under standard propagation conditions. (From Hough, M.E., *AIAA Journal of Guidance, Control, and Dynamics*, 18(5), 959, September–October 1995 [22].)

TABLE 5.2
Fixed Radar Design Parameters

Radar Parameter	Value (dB)
σ—minimum target radar cross section (m²)	−20
λ^2—radar wavelength (m²)	−20
$F_t^2 F_r^2$—minimum two-way pattern propagation factor	−20
$(4\pi)^3$—constant	33
$(S/N)_{min}$—90% probability of detection and false alarm rate 10^{-6} against nonfluctuating target	13
k—Boltzmann's constant (W s/degree)	−228.6
T_s—system noise temperature at antenna terminals (K)	30
B_n—noise bandwidth or $1/\tau$ (Hz or 1/s)	43
L—losses	6
Total (dBm⁴)	43.6
Required total (dBm⁴)	188
Required $P_t G_t G_r$ (dB)	144.4

Radar Design Parameter	Passive Array	Active Array
Transmitter peak power, P_t	1 MW	0.2 MW
Transmit antenna gain, G_t	43.2 dB	41.7
Receive antenna gain, G_r	41.2 dB	39.7
Antenna area, A	16.6 m²	11.8 m²
Antenna diameter, D	4.6 m	3.9 m
Number of elements, N	5750	4100
Peak power per element	175 W	49 W

large as a millisecond are possible, but the longer the pulse width, the longer the radar minimum range. The radar receiver cannot be turned on until after the pulse is transmitted to isolate the radar receiver during transmit. The minimum radar range is

$$R_{min} = c\tau/2 \tag{5.27}$$

where c is the speed of light (3 * 10⁸ m/s). For a 1 ms pulse width, the minimum radar range is 150 km.

Long pulse widths can be useful for searching long-range targets. The 50 μs pulse width used in the aforementioned example has a 7.5 km minimum range. An energy management design feature to account for this could be to reduce the pulse width as the targets approach the radar.

Integrating multiple radar pulses typically requires more radar time. In an unambiguous range waveform, the next pulse is transmitted after the time required for a target return from the maximum range of interest. This approach can potentially increase radar search frame times and may be more appropriate for holding a target in track once it is detected. Range ambiguous waveforms transmit successive pulses before the target returns arrive from the maximum range of interest. Multiple radar dwells at different pulse repetition frequencies (PRFs) are required to resolve the range ambiguity. In addition, blind ranges result over the target range interval of interest for a single PRF. So, multiple PRF dwells are also needed to unblind the range interval of interest. Both of these approaches can potentially increase radar search frame times and may be more appropriate for holding a target in track once it is detected to minimize radar timeline impact.

5.5.1 Radar Architecture

Radar architecture can also be exploited to achieve radar detection range requirements. An alternative radar design would be to use an active phased array approach that provides a sensitivity improvement over a passive array of approximately 10 dB. This reduces the required $P_t G_t G_r$ to 134.4 dB. Again allocating the requirement equally among the transmit power and transmit and receive gain results in a transmitter peak power requirement of 30.2 kW and an antenna size of 24 m² or a diameter of 5.5 m containing approximately 8350 elements. A T/R module at S-band can provide 50–100 W of peak power. A 0.2 MW total peak power requirement is feasible. This results in an antenna size of 9.3 m² or a diameter of 3.5 m containing approximately 3250 elements. The total transmit power can be achieved with approximately 55 W of peak power per element.

The sizing of radar antennas discussed earlier assumes that both transmit and receive aperture illuminations are uniform across the array elements. A uniform aperture weighting function provides the highest gain; however, the sidelobes are also somewhat high. A uniform circular illumination results in a first sidelobe level of approximately –17 dB. In practice, a low sidelobe weighting is used for the receive pattern to mitigate interference in the sidelobes from jamming and clutter. A weighting for receive sidelobes of –50 dB results in a taper loss, η, of approximately 2 dB. This taper loss reduces the receive gain by 2 dB. To compensate for the taper loss, the radar design must be adjusted to incorporate higher transmitter peak power, larger antenna area, or a combination of both.

Increasing the antenna area by 1 dB or 26% will compensate for the 2 dB taper loss. The resultant $P_t G_t G_r$, antenna area and diameter, number of array elements, and peak power per element summarized in Table 5.3 for notional passive and active S-band phased array designs will satisfy the detection range requirement. In the context of the system design, the detection range requirement would support missile engagements that support the keep-out volume requirement against the required targets. Table 5.3 provides a summary of S-band phased array design options.

The active array design results in a somewhat smaller aperture and has lower transmitter power requirements, potentially allowing it to be placed

TABLE 5.3
Summary of S-Band Passive and Active Radar Phased Array Designs

	Passive Array	Active Array
Peak transmit power	1 MW	0.2 MW
Transmit loss	3 dB	1 dB
Duty factor	0.02	0.10
Efficiency	0.10	0.20
Average radiated power	10 kW	15.9 kW
Prime power required	100 kW	79.4 kW
Dissipated power	90 kW	63.5 kW

higher on a shipborne platform. The increased height would provide some radar horizon extension, potentially increasing the detection range against low-flying targets.

Phased array radars, unlike older passive rotating radars, provide horizon to zenith coverage by electronically scanning the beam throughout the coverage volume. In order to provide full hemispherical coverage, phased array radars usually consist of three or four identical array faces that are equally spaced in azimuth. In active phased arrays, the transmitter and receiver functions are also contained in the arrays. In advanced active phased arrays, some of the signal-processing functions are also in the antenna. These factors tend to make the cost of the phased array system much more expensive than a rotating radar system. In some cases, rotating phased arrays are used that scan the beam electronically in elevation while covering azimuth with the mechanical rotation.

The multiface phased array radars can typically scan the coverage volume at higher rates and have the capability to electronically tailor the scans so that target sectors have higher revisit rates. Typically, the coverage volume is divided into a short-range horizon sector and a long-range above horizon sector. The horizon sector has much shorter revisit times usually on the order of several seconds. In addition, if the radar track loading in the horizon sector becomes high, the revisit rate in the above horizon sector can be slowed to maintain the horizon search frame time or to minimize the increase in horizon search frame time.

5.5.2 Platform Constraints

The platform needs to provide prime power and cooling for the radar system. In addition, the weight of the radar antenna will limit how high it can be placed in a shipborne application. The prime power requirement must account for the efficiency of a radar system. The efficiency is defined as the ratio of power radiated to prime power required. A passive phased array radar typically has an efficiency of 10%, while active phased array radar can have an efficiency of up to 20%. The radiated power is the average transmit

TABLE 5.4

Radar Power Requirement Summary

	Passive Array	Active Array
Peak transmit power	1 MW	0.2 MW
Transmit loss	3 dB	1 dB
Duty factor	0.02	0.10
Efficiency	0.10	0.20
Average radiated power	10 kW	15.9 kW
Prime power required	100 kW	79.4 kW
Dissipated power	90 kW	63.5 kW

power reduced by the transmit losses. The average transmit power is the peak transmit power times the duty factor. Typical duty factors for passive and active phased array radars are 2% and 20%, respectively. Transmit losses are typically about 3 dB for a passive phased array and 1 dB for an active phased array. The prime power requirements for the passive and active array designs are derived using the data summarized in Table 5.4.

The passive phased array radar requires 100 kW of prime power compared to 79.4 kW of prime power for the active phased array radar, which means that the active phased array radar required approximately 21% less power. The dissipated power is directly correlated with cooling requirements. The active array radar requires approximately 30% less cooling capacity to remove the dissipated power heat load.

The weight of an active phased array antenna will typically be greater than a passive phased array antenna for a fixed antenna diameter. The active array antenna will be heavier due to the distributed receiver and transmitter components that are integrated into the array face. For the passive and active array radar designs, the antenna diameters are 4.1 and 3.5 m, respectively. Since arrays are not the same diameter, a detailed weight analysis would be required to determine actual antenna weights. One way to potentially reduce the antenna size and weight is to change the radar frequency from S-band to X-band. A given aperture size will produce about 5 dB more antenna gain at X-band versus S-band. Therefore, the two-way gain is increased by about 10 dB. This allows smaller and lighter-weight antennas to be considered; however, changing the frequency band has other performance considerations that were discussed in the requirements section.

5.6 Engagement System Performance Requirements

The engagement system (ES) is required to destroy the target system (TS). The engagement system operates when supported by the central defense system (CDS) as specified by the system performance specifications. From this point forward, the engagement system will be limited to air defense missile systems.

The first air defense interceptor (ADI) performance requirement to address is the engagement boundaries necessary to defend the keep-out zone of the defended asset. This is necessary before the engagement system requirement development process can be put into place. A notional range–altitude engagement boundary can be derived as shown in Figure 5.19. Depending on the target characteristics, the engagement boundaries for a given ADI will change. High-speed targets will, for example, collapse the engagement boundaries more quickly than low-speed targets due to the fact that the success of the engagement is dependent on not breeching some minimum dynamic pressure to sustain maneuverability and time-constant performance and having a minimum homing time. Therefore, any set of engagement boundaries must assume that target acquisition occurs at some minimum kinematic boundary of the missile system that statistically ensures an acceptable P_{ssk}. Range–altitude boundaries will decrease in areas when the targets are in an ECM

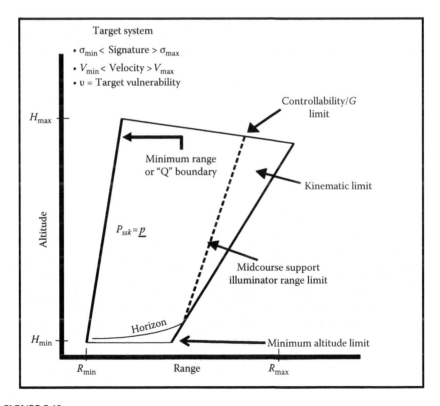

FIGURE 5.19
Engagement boundary requirements.

environment, in a high sea state (when over water), against a maneuvering target or when a number of other anomalous conditions exist. Therefore, the engagement boundary requirement is specified with a number of qualifiers including specific performance conditions. In the capability-based acquisition process, a required engagement boundary is developed but it may need to be *purchased* over many iterations/years in a spiral fashion. It is important to first deploy some measurable capability with a system and plan for growth in achievable steps that will satisfy both fiscal and schedule realities.

Figure 5.19, range–altitude engagement requirements, displays performance-limiting factors. These factors need to be considered in the engagement system requirement flow down process. Initial engagement boundary design criteria include minimum acceptable P_{ssk}, which incorporates interceptor miss distance criteria and target signature, velocity profile, and vulnerability characteristics. The requirement shown in Figure 5.19 can be interpreted as that performance when launched on a standard atmospheric day while engaging a nonmaneuvering benign target described by the associated features (σ, V, υ). The missile is required to be capable of intercepting targets with a minimum $P_{ssk} \geq p$ at an altitude greater than or equal to H_{min} meters and an altitude of less than or equal to H_{max} kilometers and within a range boundary of no less than R_{min} kilometers and out to a maximum range R_{max} kilometers. The minimum and maximum range–altitude boundary will exist due to minimum dynamic pressure and thus maneuverability and kinematic limits. Achieving acceptable outer and inner range–altitude boundaries will be the focus of design driven primarily by target kinematic and dynamic capabilities. Figure 5.19 also shows the maximum boundary where the CDS will be required to handover guidance and control to the engagement system. If handover is delayed to this point in the envelope, then an active or passive sensor system or suite will need to be employed by the engagement system. Were handover to occur prior to this limit, then a semi-active radar sensor system would fulfill requirements and may be preferable. Essentially, this limit is the horizon or the electronic horizon that will vary due to anomalous propagation conditions and frequencies employed.

The next boundary that will need to be defined is the Mach–altitude envelope. The trade space of the AMD system will ultimately measure achieving acceptable P_{ssk} beyond the keep-out zone, which in turn is a time requirement. The time requirement will be satisfied by balancing the sensor suite detection capability against the engagement system flyout time contour. Cueing from ISR components and *speeding up* CDS reaction time will play a role in the trade space but will be shown to have second-order effects in comparison to achieving increased detection range and speed at which ordnance can be delivered. ISR cueing and reaction time reduction will be fine-tuning, whereas the sensor suite detection range capability and the engagement system speed will be critical design parameters.

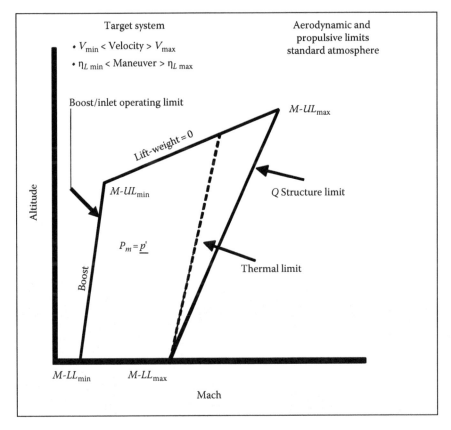

FIGURE 5.20
Engagement Mach–altitude envelope. (Modified from Menon, P.K.A. and Briggs, M.M., *AIAA Journal of Guidance, Control, and Dynamics*, 13(4), 596, July–August 1990 [24].)

A notional Mach–altitude engagement envelope is shown in Figure 5.20 that displays performance-limiting factors that need to be considered in the aerodynamic–propulsive design requirement flow down process. The engagement boundary Figures 5.19 and 5.20 and the itemized requirements are linked by altitude and P_{ssk} through miss distance requirements. Miss distance is tightly coupled to missile time constant that is in turn tightly coupled to dynamic pressure (Q), or velocity. There are four requirements specified in Figure 5.20 that need to be achieved. M-LL represents the two Mach–low-altitude limit requirements and M-UL represents the two Mach–upper-altitude limit requirements. The requirement driver achieves a minimum miss distance probability, $P_m \geq p'$. Beyond this requirement set, the limiting factors are airframe/radome thermal, structural, and lift-to-weight ratio limits. The Mach–altitude envelope requirement will need to be expanded to deal with a more energetic target set. The primary factors to consider are

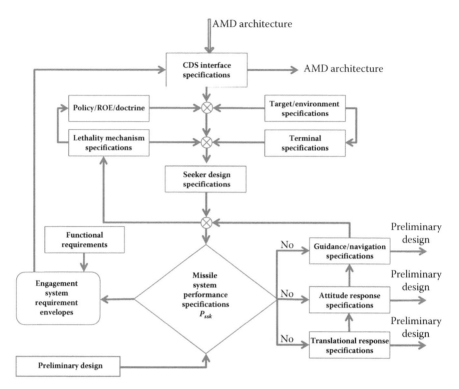

FIGURE 5.21
Engagement system specification development process.

target speed and maneuverability limits. As these two target characteristics increase, the Mach–altitude boundary requirements will need to grow. A boost propulsion system is required to achieve a minimal Mach limit where a sustain propulsion system can take over. The sustainer can be either a solid rocket motor or an air-breathing propulsion system. An air-breathing system will involve additional or more stringent limiting factors that need to be considered to meet the Mach–altitude requirement including angle of attack, propulsion chattering, and inlet operating limits.

One last note on engagement boundaries is necessary. The engagement boundaries need to represent the acceptable degradation in performance due to conditions to be specified in the remaining part of this process.

Once the engagement system requirement envelopes are developed and the blanks that need to be determined are defined, it is then possible to begin assembling the specification tree that will completely identify the system performance and interface requirements. Specifications are the documentation of requirements that will include identifying what the system is being designed to accomplish (the system function). An important element of the *function specifications* is to document the purpose for the system. This

Phase A: AMD System Requirements

is accomplished through the engagement boundaries, Figures 5.19 and 5.20, derived from a flow down of the functional requirements.

Next, the components make up the system and how those components interrelate (interface requirements) is established. This is also sometimes referred to as the budgeting and allocation process. The *system budget* will include a breakout of the required operations of the system and an allocation to subsystems, components, or elements. Specific rules on how to use the system in conjunction with operations and components to meet the system requirements are all essential parts of the budgeting process. Finally, this process will include identifying how the system and its components will perform (performance requirements), which includes defining how the performance is to be verified (inspection, test, or demonstration). A performance margin must be established to ensure that both design and system meet the requirements.

Figure 5.21 shows the flow for engagement system specification development. This begins with the functional requirement flow down into the formation of engagement envelopes discussed earlier. The complete set of CDS interface specifications must flow down. CDS interface specifications have also evolved from and are evolving from the specification development process within the CDS that includes sensor suite, battle management, and engagement computer systems. In other words, the development of system-level requirements and subsequently the specifications will be part of the trade space that needs to flex with the preliminary design development in order to achieve mission goals and within the other constraints.

The process requires a first pass to define five specifications before proceeding further. The policy/rules of engagement/doctrine is an additional set of constraints to be placed on the development of engagement solutions. This includes prohibited cost boundaries, prohibited solution sets (at one time, e.g., there were prohibitions from developing space-based interceptors), minimum acceptable cumulative probability-of-kill solutions, prohibited or preferred kill mechanisms (this could include nuclear or other types of politically sensitive approaches), and launching conditions to include sea, land, space, air environments, and specific launching systems. The extremes of launching conditions will be specified as well to include weather conditions (i.e., temperature, wind precipitation extremes).

The target and environment specification will be developed to include whether engagement specifications are in a clear environment, electronic attack environment, multiple target environments, or possibly a number of other conditions.

The constraints combined with the target set will in turn drive/limit the remaining three first pass specifications: lethality mechanism, seeker, and terminal approach specifications. This loop may take several iterations to develop a comprehensive set of requirements and specifications that can then flow down to the remaining pieces of the process.

Lethality mechanism, seeker, and terminal approach strategy are closely coupled specifications. A hit-to-kill lethality strategy, for example, will put small wavelength constraints on the seeker system specification to permit the accuracy that will be required to ensure the required P_{ssk}. Moreover, the targets' characteristics will drive which wavelength strategies may be preferred. The terminal homing approach may be dictated as the combination of constraints accumulates. Midcourse guidance performance provided by CDS will drive seeker requirements and design decisions.

5.7 System Requirements Document

Once the specifications are settled for the first complete pass through the process and the P_{ssk} performance requirement has been settled relative to the required engagement boundaries, which may be a variable mapping throughout the boundaries, it is necessary to document the conclusions in the system requirements document.

The SRD completes Phase A of the systems engineering process with a review by the entire acquisition team. The document should contain the completed work including the mission needs statement, TLR/MOEs, CONOPS, top-level architecture (TLA), and system specifications. Within the TLA, there will be an air and missile defense system, target system, and ISR system. Within the air and missile defense system, there will be the central defense system and the engagement system. Within the CDS, there will be a sensor suite system, battle management system, and engagement computer system. System specifications are developed and flow through the architecture beginning with the establishment of a technical and policy based keep-out volume with respect to defended assets.

The ISR system specifications include producing the target origin, conducting geopositioning, and communicating to the AMD system. The AMD system specifications include producing the required engagement boundaries, detect and track, doctrine, engagement solutions, and communication throughout the AMD system. Assembling detailed target characterizations with respect to sensor suite system and engagement system solutions as specifications are also required.

Moreover, it is proposed that within the engagement system the constraints that apply to AMD engagement be imposed. Although there will be subordinated constraints associated specifically with their operation within the other systems, it is within the engagement system operation that the constraints will flow more effectively and efficiently. Engagement system requirements include flyout Mach–altitude and range–altitude boundaries, midcourse guidance, terminal homing, and lethality strategies. The engagement system communication requirements are specified within the air and missile defense system.

References

1. Foard, T.R., Theater air defense cornerstones, *Johns Hopkins APL Technical Digest*, 21(3), 387–392, 2000.
2. Yopp, S.W., *Aegis and the C/JFACC: The Naval Surface Combatant as an Element of Air and Space Power*, AU/ACSC/2395/AY06, Air Command and Staff College Air University, Montgomery, AL, April 2006.
3. Hartwig, G., Aegis, from MIL-SPEC to open systems, in *NDIA Systems Engineering Conference*, Tampa, FL, October 21–24, 2002.
4. O'Rourke, R., Sea-Based Ballistic Missile Defense—Background and Issues for Congress, RL33745, CRS Report for Congress, Washington, DC, November 21, 2008.
5. Lawes, I.S., Defence against terrorism—A role for ground based air defence, in *Land Warfare Conference 2007*, Adelaide, South Australia, Australia, October 2007.
6. Lussier, F. et al., *Army Air and Missile Defense-Future Challenges*, RAND Corporation, Santa Monica, CA, 2002.
7. GAO-04-175, *Nonproliferation—Improvements Needed to Better Control Technology Exports for Cruise Missiles and Unmanned Aerial Vehicles*, Washington, DC, January 2004.
8. Weidler, R.D., Target LOS and seeker head equations for an AAM simulation, Flight Dynamics Group, Stability and Flight Control Section, FZE-977, November 19, 1969.
9. Bose, S.C., Radar updated strapdown inertial midcourse guidance performance analysis for missiles, AIAA Paper 79-1726, Reston, VA, 1979.
10. Kouba, J.T. and Bose, S.C., Terminal seeker pointing-angle error at target acquisition, *IEEE Transactions on Aerospace and Electronic Systems*, AES-16(3), 313–319, May 1980.
11. Stallard, D.V., Classical and modern guidance of homing interceptors, Seminar of Department of Aeronautics and Astronautics, Massachusetts Institute of Technology, Cambridge, MA, April 1968.
12. Lin, C.F., *Modern Navigation, Guidance and Control Processing*, Prentice Hall, Englewood Cliffs, NJ, 1991.
13. Serakos, D. and Lin, C.-F., Linearized kappa guidance, *Journal of Guidance, Control, and Dynamics*, 18(5), 975–980, 1995.
14. Lee, G.K.F., Estimation of the time-to-go parameter for air-to-air missiles, *AIAA Journal of Guidance, Control, and Dynamics*, 8(2), 262–266, 1984.
15. Massoumnia, M., Optimal midcourse guidance law for fixed propulsive maneuvers, *AIAA Journal of Guidance, Control, and Dynamics*, 18(3), 465–470, May–June 1995.
16. Hablani, H., Pulsed guidance of exo-atmospheric interceptors with image processing delays in angle measurements, in *AIAA Guidance, Navigation and Control Conference and Exhibit*, Denver, CO, August 14–17, 2000.
17. Ben Asher, J.Z. and Yaesh, I., Optimal guidance with a single uncertain time lag, *AIAA Journal of Guidance, Control, and Dynamics*, 18(5), 981–988, September–October 1995.

18. Kumar, R.R., Seywald, H., and Cliff, E.M., Near-optimal three-dimensional air-to-air missile guidance against maneuvering target, *AIAA Journal of Guidance, Control, and Dynamics*, 18(3), 457–464, May–June 1995.
19. Kumar, R.R., Seywald, H., Cliff, E.M., and Kelley, H.J., Three-dimensional air-to-air missile trajectory shaping, *AIAA Journal of Guidance, Control, and Dynamics*, 18(3), 449–456, May–June 1995.
20. Bezick, S., Rusnak, I., and Gray, W.S., Guidance of a homing missile via nonlinear geometric control methods, *AIAA Journal of Guidance, Control, and Dynamics*, 18(3), 441–448, May–June 1995.
21. Yang, S.-M., Analysis of optimal midcourse guidance law, *IEEE Transaction on Aerospace and Electronic Systems*, 32(1), 419–425, January 1996.
22. Hough, M.E., Optimal guidance and nonlinear estimation for interception of accelerating targets, *AIAA Journal of Guidance, Control, and Dynamics*, 18(5), 959–968, September–October 1995.
23. Imado, F., Kuroda, T., and Miwa, S., Optimal midcourse guidance for medium-range air-to-air missiles, *AIAA Journal of Guidance, Control, and Dynamics*, 13(4), 603–608, July–August 1990.
24. Menon, P.K.A. and Briggs, M.M., Near-optimal midcourse guidance for air-to-air missiles, *AIAA Journal of Guidance, Control, and Dynamics*, 13(4), 596–602, July–August 1990.
25. Yanushevsky, R., *Modern Missile Guidance*, CRC Press, Boca Raton, FL, 2008.
26. Blake, L., *Radar Range-Performance Analysis*, Munro Publishing Company, Silver Spring, MD, 1991, p. 17.
27. Courtesy of Huong Pham of Technology Service Corporation, Silver Spring, MD.
28. AIAA/ANSI, R-004-1992, *Recommended Practice for Atmospheric and Space Flight Vehicle Coordinate Systems*. AIAA, 370 L'Enfant Promenade, SW, Washington DC 20024, 28 February 1992.
29. Sharp, E., A triangular array of planar-array elements that reduces the number needed, *IRE Transactions on Antennas and Propagation*, 9, 126–129, March 1961.

6

Phase B: Preliminary Design

Air and missile defense (AMD) systems' preliminary design requires a complex and interactive set of design, modeling, and simulation processes and toolboxes that capture functional, performance, and interface requirements [1–7,21–23]. The Battlespace Engineering Assessment Tool (BEAT) was developed by the authors to conduct (1) sophisticated ship combat systems engineering preliminary design; (2) requirements trade studies; and (3) war-fighting performance analysis. An analysis of BEAT will enable a complete discussion of the AMD preliminary design process. For example, BEAT was initially used for both cruise and ballistic missile defense applications. Specifically, BEAT was used to quantify battlespace performance for complex air and missile defense mission areas such as ship self-defense (SSD), area air defense (AAD), and ballistic missile defense (BMD) consisting of several modeling and simulation toolboxes linked together with specialty software and engineering processes. Within each toolbox, there are multiple models and simulations employed. For the purposes of this book, only the BEAT AMD systems engineering preliminary design process will be discussed. It is important to note that BEAT is not a turnkey system but more generally represents an engineering process for dealing with current air and missile defense system problems. Attempting to develop a "turnkey" AMD systems engineering preliminary design tool is unadvisable. Problems and decisions encountered in this process need assessment and evaluation at each step that can only be evaluated by a skilled engineering team.

Specific models, simulations, and analytical tools within BEAT are employed based on the problems being addressed and the questions that need to be answered. Tools, models, and simulation will come and go as part of the process. There is no one tool for any one of these processes that will accurately address the matrix of alternatives and engineering problems that will be encountered. The battlespace engineering process and BEAT are shown in Figure 6.1. Results from Figure 6.15 used to develop radar design options and Figure 6.19 used to develop interceptor design options are inputs to the BEAT process shown in Figure 6.1.

94   *Air and Missile Defense Systems Engineering*

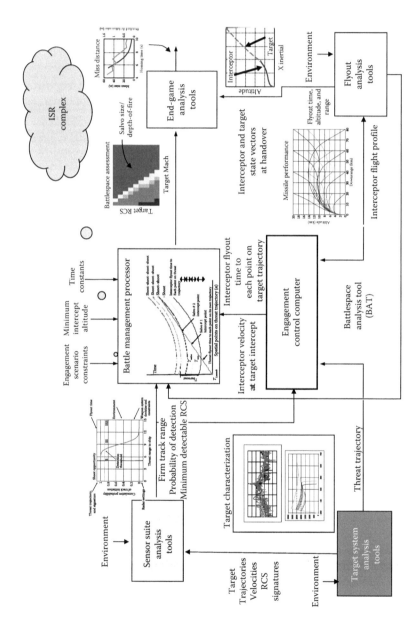

FIGURE 6.1
Battlespace Engineering Assessment Tool.

6.1 Target System

The target system tool (TST) box contains specific and general models that are used as part of an analytic process. Target missile engineering characterizations, models and simulations, and/or database such as detailed trajectories, radio frequency (RF), infra-red (IR) and imaging infrared (IIR) signatures, other important dynamic features, and analytical tools to time/space correlated trajectories and signatures are part of this toolbox. Characterization fidelity is tailored to match the specific problem being studied. In some cases, it may only be required to, for example, use a single-number radar cross section (RCS) and define the associated Swerling number(s). In other instances, time/space correlated full aspect, polarization, and frequency-dependent signatures are needed.

6.2 Sensor Suite

The sensor suite specification process is shown in Figure 6.2. The sensor suite design options are determined in the process shown in Figure 6.2 and represented in the sensor suite tool (SST) box in Figure 6.1. Results from Figure 6.1 process feed back into Figure 6.15 and complete the cycle. SST requires detailed modeling capabilities to simulate potential radar systems with physics-based specifications from Figure 6.2 and potential design variations including various multifunctional tracking modes and signal processing techniques.

Antenna specifications will include the horizon and above horizon search lattice design, off-broadside and boresight angle design limits, and beam search mode operation requirements. The transmitter specifications will include waveform design options as a function of environment and target considerations, transmitter dwell timing design, and beam search mode design option requirements. Receiver specifications will include signal processing design requirement options. Specific modes will need to include clear, anomalous environment and jamming modes. Moving target indicator (MTI) or pulse Doppler (PD) modes will be essential to discriminate targets against land mass backgrounds and in certain jamming environments. Sensitivity time control (STC) design requirements will be driven by a variety of factors that include environment and target considerations. Finally, the sensor suite may be required to provide data and/or commands to the missile in flight. Engagement system communication specifications will be developed here. Elements that will be considered will depend on engagement system CDS communication requirements. Whether or not a missile rear reference receiver is required to synchronize the missile seeker intermediate frequency (IF) for Doppler processing will need to be determined

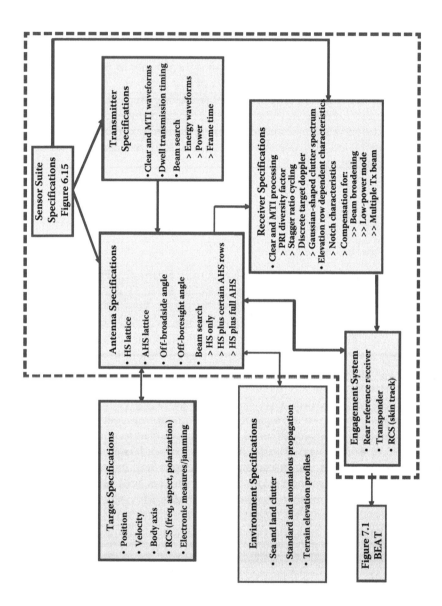

FIGURE 6.2
Sensor suite specifications and preliminary design process.

Phase B: Preliminary Design

iteratively. Whether or not the engagement system will need to communicate back to the CDS may require a transponder link. The performance of these elements will affect the engagement.

Illuminator and illuminator schedule function specifications if appropriate are also located in the sensor suite tool set.

6.3 Battlespace Assessment

Figure 6.1 shows the battlespace assessment tool (BAT) in yellow, and it has two specific analytical tools, the battle management processor (BMP) and the engagement control computer (ECC), to emulate the CDS functionality. These two tool sets allow a balanced systems engineering approach to be applied to the AMD preliminary design process. The BMP contains algorithms and logic (developed by Slack [12]) that will be discussed in the following paragraphs of this chapter. The BMP algorithms and techniques provided here will permit battlespace analysis to be conducted regardless of the target scenario, trajectory, or a multitude of other potential engagement variables. Figure 6.3 is an illustration of a set of engagement scenarios for an illustrative AMD system. The orange lines represent imaginary rays from a sensor suite that make up a radar fence for sequential AMD system positions. The ballistic target has to fly through the fence to impact the desired impact locations. The target trajectories may represent an illustrative ballistic

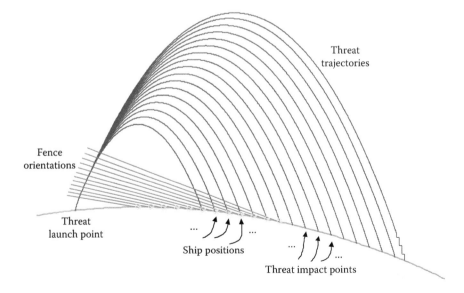

FIGURE 6.3
Illustrative BMD engagement scenarios.

target missile with a range uncertainty shown by the trajectory variations, or they may represent a number of target missiles whose ranges span the extent demonstrated in the graphic and represents the potential uncertainty facing the AMD system.

Figure 6.4 illustrates the complex nature of any single engagement of a ballistic target that must take place in a timeline. The parabolic lines showing movement from left to right represent the target missile, and the defensive interceptor flight path events that take place during the engagement. The events are presented in text relative to ground range and in time.

Thresholds are shown as straight horizontal lines including the minimum altitude at which an intercept can occur and the earliest altitude at which the interceptor sensor becomes operational. All other events are self-explanatory. Each of these events represents a part of the trade space that needs to be studied during requirement development and the preliminary design process. The uncertainties shown in Figure 6.3 will shape the trade studies. The techniques used to capture this set of trades begin with mapping all the events into a common time/space frame of reference as shown in Figure 6.5 where the target trajectory is mapped into a time history.

Figure 6.6 shows, in an illustrative manner, that if an interceptor were launched simultaneously with the target how its spatial points would map into the time/space history of the target. This can be accomplished using numeric techniques to synchronize the interceptor flyout timeline to the target flyout timeline. Where and when the timelines intersect in space represents the first opportunity for intercept. This solution does not yet take into account the additional time and space constraints that need to be considered in the design requirements. Moreover, doctrine considerations will impact the solution set. The first consideration is the time it will take the target to fly through the radar fence representing the first opportunity for detection. Next, assuming that the first detection occurs, it is then necessary to compute the battlespace timeline requirement and add that time to the mapping. Figure 6.7 shows the addition of first detection and battlespace timeline requirement for a single-shot opportunity.

Figure 6.8 shows an overlay where multiple-shot opportunities are examined and the system is determined to be limited to a dual-salvo doctrine.

Any number of constraints, operations, and events can be included in the trade space. Figure 6.9 shows how a discrimination time budget can be added to the engagement strategy. Note how it can be implemented in several ways. Figure 6.9 is constructed assuming that the intercept occurs immediately after the discrimination process is complete. This is simply an example of the budgeting process to produce an intercept solution. The engagement strategy may also include completing an intercept during discrimination and after a fixed number of discrimination seconds. The process simply requires a discrimination time budget allocation and the ability to shift the solution by the amount of time being provided for discrimination purposes. The important part of this process is that a time budget is required to begin a requirement

Phase B: Preliminary Design

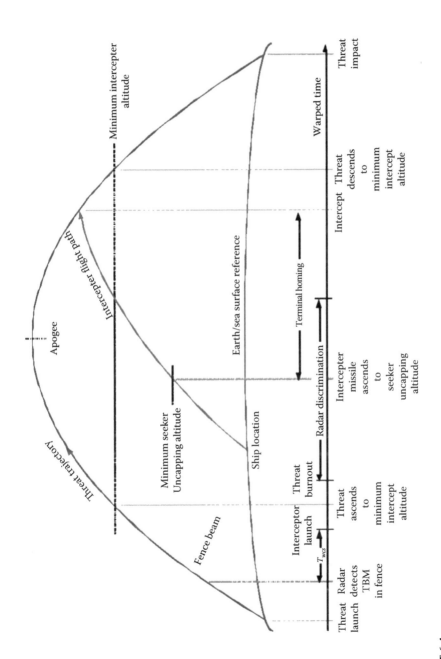

FIGURE 6.4
Ballistic missile defense battlespace timeline details.

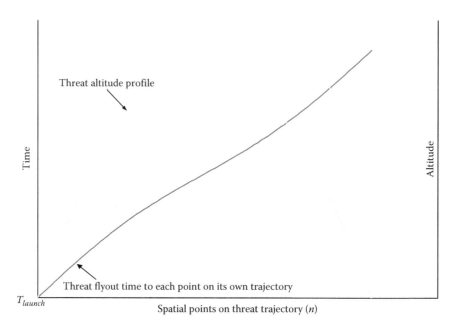

FIGURE 6.5
BMD engagement target time–space correlation.

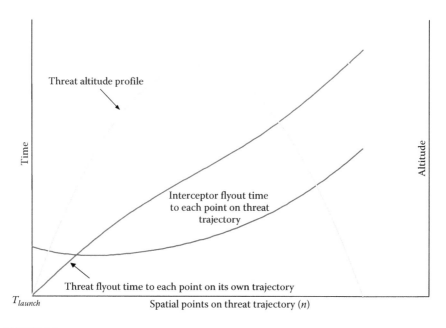

FIGURE 6.6
BMD target and interceptor time–space correlation.

Phase B: Preliminary Design

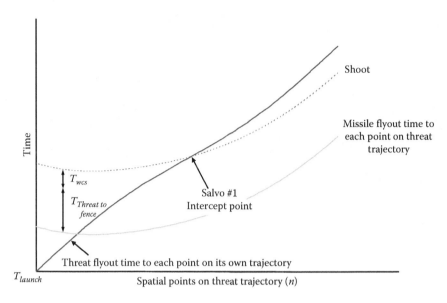

FIGURE 6.7
BMD target and interceptor single-shot opportunity.

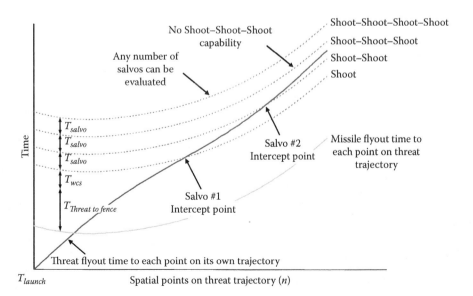

FIGURE 6.8
BMD target and interceptor multiple-shot opportunities.

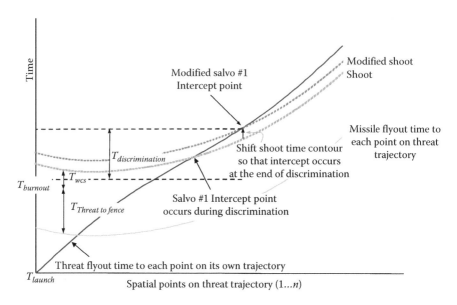

FIGURE 6.9
BMD target and interceptor including discrimination.

definition for AMD systems. Many eloquent discrimination solutions may exist, but unless the AMD system timeline budget supports their implementation, they are merely academic. Thus, it is necessary to begin design requirement development by producing an engagement timeline budget.

The engagement control computer process houses computer programs that will translate weapon engagement orders from BMS into commands for the control and management of engagements. The ECC emulates the AMD system ECC fire-control function that includes filtering sensor suite radar data, computing predicted intercept point and midcourse guidance commands, predicting missile time, and estimating time to go. The ECC can be configured to simply provide rule-based timeline budget information to the BMS tool based on inputs from the sensor suite, target system, and engagement system tool sets, or it could explicitly model these functions. The iteration state of the requirement development or design process may dictate the details necessary for the ECC configuration. As the design phase matures, the fidelity requirements on the ECC will increase.

6.4 Engagement Analysis

The purpose of engagement analysis is to produce an interceptor missile preliminary design in balance with the other elements of the AMD system.

Phase B: Preliminary Design

There are two components of engagement analysis, flyout, and end game, shown in Figure 6.1. Two sets of overlapping but different analysis tools are required to complete the engagement analysis objectives. Flyout analysis will establish the requirements when, where, how many, and which interceptor missile variants can reach the target(s). This is not the same set of requirements that will establish whether or not the interceptor missiles can destroy or *hit* the target(s). Flyout requirements are a matter of establishing interceptor missile *reach* and timeliness to ensure that multiple engagement opportunities will exist. The aggregate P_k requirement forces the trade space to include having a sufficient number of interceptors (usually greater than 1) reach the target or have it met with one interceptor. The latter is a tall order and is not the likely outcome of the trade-space analysis.

Reach performance is likely established (when, where, how many, and which variants can reach the target[s]) after the *end-game* performance requirements are produced and the preliminary design iteration is completed. *End-game* analysis determines whether or not a target kill can be achieved given a specific *reach* performance. It is advisable to back out the requirements beginning at the point where intercept is desired and with the knowledge of what terminal conditions will consummate a *kill* given a specific lethality strategy. Once the *kill* criterion is established, terminal homing requirements should follow that dictate handover requirements. Parametrically determine what the handover requirements must be to achieve terminal homing that satisfies the *kill* criterion and so on backing up into the beginning of the kill chain of events, establishing the performance of those events and the time budget requirements for those events. At the other end of the engagement will be the engagement envelope requirement that must also be met. This will likely force an iteration of this design loop until both *kill* and engagement envelope requirements are simultaneously met.

Engagement envelope requirements will flow down from the top-level requirement (TLR) process and specifically from Figures 5.19 through 5.21. A flyout analysis tool set will need to be developed that employs the interceptor flyout model(s) and interfaces with the TST models and midcourse guidance models from the ECC. An interceptor missile flyout model and simulation block diagram capable of supporting this phase of requirement development are shown in Figure 6.10.

There are four primary components of an interceptor missile engagement simulation, which need to be described not including the target. They are engagement physics; seeker; guidance, navigation, and control (GNC); and interceptor kinematics and dynamics. Missile interceptor flyout modeling will require a complete propulsion stack-up, aerodynamics, an accurate earth model, midcourse and terminal homing phase guidance, a representative target characterization, and a representation of the CDS uplink/downlink if one exists. Midcourse guidance instruction originates from the engagement control computer (ECC) that may be, but not necessarily, an integral part of the flyout tool set.

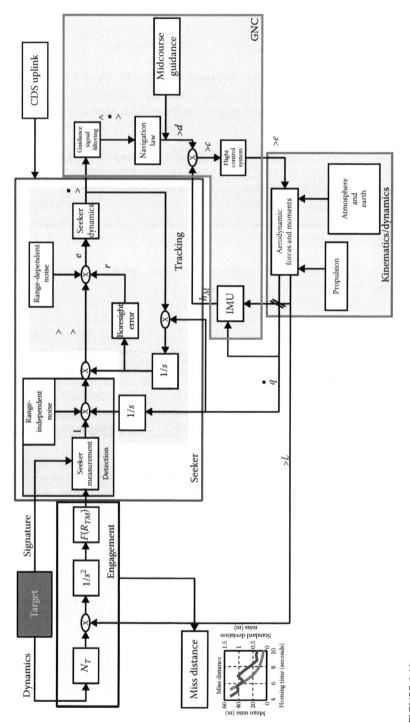

FIGURE 6.10
Simplified interceptor block diagram.

Engagement physics represents the dynamics between the interceptor missile and the target in time and space. The missile seeker observes and tracks the dynamics between the interceptor and the target and generates error signals for the GNC system to process and produces steering signals. The seeker contains detection and tracking logic and algorithms that process target reflected or direct energy corrupted by range-dependent and range-independent noise and parasitic, such as radome boresight slope, error sources. See [1–3,10,20,29,34–38,59–63 and 76] for more details on the GNC processes regarding the rest of this section.

The GNC system processes target tracking error signals first through a guidance computer to generate acceleration commands that are acted upon by the flight control system that generates steering signals for actuation. Interceptor missile kinematics and dynamics are represented by the components that produce forces and moments during flight that include interceptor aerodynamics, propulsion, and the effects on flight from the atmosphere and earth (the flight environment model). These force and moment components produce translational accelerations (kinematics) and rotational motion (dynamics).

A separate terminal end-game model to determine miss distance may be developed as shown in Figure 6.10 or Monte Carlo techniques may be used in the end-game portion of the engagement including a variety of noise sources. The end-game simulation will initialize using a deterministic flyout simulation to handover. Monte Carlo techniques will certainly need to be used to produce accurate miss distance results. The miss distance results are then used to conduct lethality analysis. Typically, an end-game model is a fully functional six-degree-of-freedom Monte Carlo simulation. Six degrees of freedom refers to modeling the three translational and three rotational motions about the translational axes. A Monte Carlo simulation is a deterministic model that is iterated numerous times to capture the potential statistical variation of a set of variables that either are stochastic in nature or have a bounded uncertainty with likely probability distribution functions. Identifying the appropriate variables for Monte Carlo variation is an essential activity in the preliminary design process. Design variables for Monte Carlo may include aerodynamic coefficients, mass and inertia properties, flight control design, and propulsion design parameters. Producing an accurate Monte Carlo simulation for missile end-game analysis also relies on identifying appropriate noise sources with realistic power spectral density characteristic functions. The simulation must also therefore include a sophisticated seeker model including detection and tracking signal processing. The flight control system dynamics, command limiting and other nonlinearities, and detailed time–space correlated target signature and flight dynamic characterizations are also essential. See [52–56] for further details on missile flight mechanics. Homing guidance laws and discrete processing must also be algorithmically captured accurately.

The design process is fundamentally tied to the modeling and simulation of the problem and elements relying heavily on affordable hardware and software capability. Computer processing power is abundant and relatively inexpensive, and the same assessment can be made of software application. Therefore, except in the concept phase, less accurate approaches to determining miss distance probability functions are not necessary.

Another distinguishing set of factors in end-game simulation analysis are target characterization requirements. *End-game* target modeling and simulation must include time-dependent target dynamic characterizations with high-resolution time–space correlated signatures. Interceptor missile terminal homing modeling must include all of the error sources associated with handover (heading, cross-range, seeker pointing angle), terminal sensor range-dependent and range-independent noise, parasitic noise, and guidance and navigation instrument (inertial reference unit [IRU], inertial measurement unit [IMU], inertial navigation system [INS], global positioning system [GPS]) noise. Error modeling has to include representative statistical distributions [1]. Lethality assessment is included in *end-game* analysis and will follow mapping the probability of achieving a miss distance criteria (P_m). Kill mechanism and target vulnerability modeling are essential in producing a lethality assessment and is conducted with yet another set of analytical tools, models, and simulations that when combined with P_m maps produce P_{ssk} maps.

The purpose of the flyout analysis tool set is to size the interceptor missile, design midcourse guidance, and further develop a terminal homing strategy to achieve the *reach* performance specified by Figures 6.19 and 6.20 (the TLR engagement boundary) with sufficient energy margin to achieve the specified P_{ssk} mapping within the TLR boundary. One way to measure this objective is to achieve a deterministic miss distance within the flyout tool set.

For the purposes of *reach* analysis, the interceptor flyout model will need to produce accurate flyout contours based on accurate aerodynamic, mass, and inertia properties, propulsion and environment modeling, numerically capturing the midcourse guidance, terminal homing, and approach angle control laws [6]. Monte Carlo techniques will be necessary to capture handover error. Once handover error can be characterized with a statistical distribution, Monte Carlo techniques are not necessary to establish *reach* analysis.

At this point, it is worthwhile to revisit Figure 5.21 to facilitate moving into the discussions on preliminary design. Prior to beginning the engagement analysis, the first preliminary design iteration will have to be developed. As depicted in Figure 5.21, terminal homing and seeker design requirements are developed and combined with the other requirements from the top of Figure 5.21 to flow down into guidance/navigation, attitude, and translational response preliminary design requirements. The preliminary design process is iterative and should begin with satisfying the seeker requirements necessary to achieve end-game performance requirements. Translational, attitude response, and guidance/navigation requirement development will follow. The next sections will follow the sequence of preliminary design and begin with interceptor missile seeker preliminary design.

6.5 Missile Subsystem Preliminary Design

6.5.1 Missile Seeker Preliminary Design

The missile seeker measures target angle only if a passive (electro-optical or ARH) sensor is employed. If an active or semi-active radar sensor is used, in addition to angle, target range, range rate, and possibly velocity are measured discriminates relative to the sensor frame of reference. An active sensor transmits and receives the electromagnetic spectrum from the same antenna in a similar fashion to the fire-control radar. Semi-active radar (SAR) seekers require target illumination from separate and distinct radars or illuminators. The receiver is on board the interceptor, and a rear reference signal must be provided to track the direct illumination signal. Typically, SAR seekers are continuous wave (CW) systems. CW systems permit target relative Doppler (velocity) and angle measurements but not range tracking.

Figure 6.11 shows how target mission and homing time requirements will drive the seeker design trade space. The chart set assumes illustrative interceptor missile terminal Mach and target average closing speeds (mission) during terminal homing. The charts extend to 6.5 seconds homing time. In some design cases, longer homing times may be desirable to minimize miss and relax interceptor responsiveness, but the trade here is the longer the homing time, the more vulnerable the seeker becomes to countermeasures.

The obvious fact here is that the fast targets stress homing time requirements and drive the interceptor design to more expensive, higher-power (or more sensitive) seeker design solutions. High-speed, highly maneuverable targets drive up homing time requirements while jamming/deception and signature reduction reduces acquisition ranges. Optimally combining these attributes are the target designer's objectives to defeat the interceptor. Optimizing minimum seeker homing time and signal processing strategy while reducing interceptor attitude response times to improve P_{ssk} is the interceptor designer's objective. Low-altitude targets present a significant environmental problem with clutter and multipath/reflections that drive up homing time requirements [2,3,20,61,62]. Homing strategy and time requirements are also driven by kill criteria in addition to interceptor attitude response design. Unfortunately, there are no closed-form solutions to solving all of these trades simultaneously, and achieving a successful design will require iteration. The SRBM defense mission (shown as Mach 7 Terminal Target Mach number) is considerably different than the low-slow cruise missile defense mission (shown as Mach 0.8 Terminal Target Mach number), from both a target and environment aspect, and will likely result in two different missile seeker designs for optimum performance. Some of the seeker strategy trades will be briefly examined.

A semi-active radar-only missile is limited to the shooter line of sight or is dependent on an inorganic illuminating source. Neither option is attractive in modern missile warfare. A passive optical or infrared strategy is highly susceptible to atmospheric conditions and is limited to short acquisition ranges and homing times. This may be an option at extremely high or exoatmospheric altitudes. A passive RF strategy is dependent on the target actively emitting energy. An active radar-only missile will place more stressful requirements on the CDS and handover due to the shorter acquisition ranges and minimized homing time. Dual-mode systems, where combining these techniques to exploit various portions of the

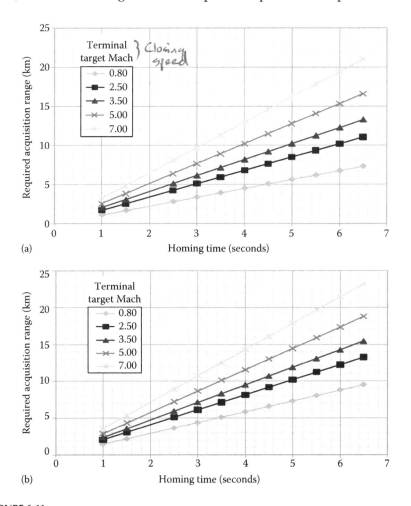

FIGURE 6.11
(a) Target impact on terminal homing trade space for Mach 2.5 interceptor. (b) Target impact on terminal homing trade space for Mach 3.5 interceptor. (*Continued*)

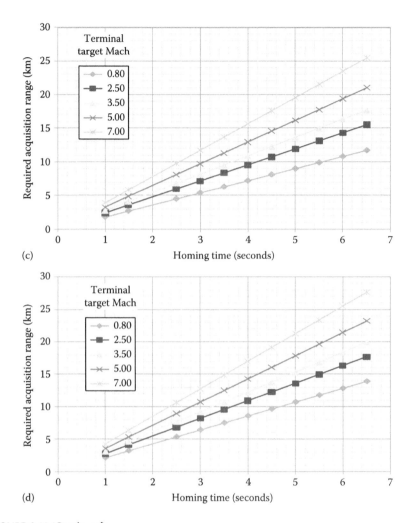

FIGURE 6.11 (*Continued*)
(c) Target impact on terminal homing trade space for Mach 4.5 interceptor. (d) Target impact on terminal homing trade space for Mach 5.5 interceptor.

spectrum as a function of target design and environmental conditions, are likely to be the most practical seeker design strategy approach.

As neither time nor space will permit a presentation of an exhaustive preliminary design trade-space study examining all of the seeker variant options, it will be assumed this was accomplished and Section 6.5.1 will focus on the design and performance trade-offs of an active radar seeker design for illustrative purposes. The generality of the preliminary design approach holds regardless of the specific sensor type chosen.

6.5.1.1 Angle Tracking

The interceptor missile seeker is required to provide highly accurate target angle location and angular time derivatives at high angular LOS rates while isolating the sensor from missile body motion. The two design objectives are angle resolution and angle measurement accuracy. To accomplish angle tracking, two processes are required: target spectrum signal collection and signal processing. Signal collection can be accomplished with either the electromagnetic or electro-optical spectrum. Depending on the target and the background environment, one approach may be preferred over the other. Regardless of which part of the spectrum is used, the signal received at the interceptor missile contains target and noise information within a volume. This is referred to as the received space–time correlated signal (S_r) that can be represented by the following equation (Maksimov and Gorgonov [21], Chapter 6):

$$S_r(\tau, \rho) = S_t(\tau, \rho, \xi) + S_n(\tau, \rho) \tag{6.1}$$

The signal (S) subscripts r refers to received, t refers to target, and n refers to noise. The independent functional variables τ refers to time, ρ refers to the radius within the volume being detected by the signal, and ξ refers to the desired information vector content (LOS and LOS rate). Signal processing (temporal and spatial) is applied to S_r to resolve the angular properties of the target signal without the noise. This process is referred to as target angular discrimination. The apparatus used to collect the microwave signal spectrum is an antenna and to collect the infrared spectrum is optics.

There are many types of antennas that can be used to fulfill the spatial filtering, angle tracking, requirement including, for example, Cassegrain twist and planar phased arrays (see James [20], Chapter 4). The most practical solution to meet interceptor missile antenna requirements is the mechanically scanned slotted planar phased array (James [20], Chapter 4, and Maksimov and Gorgonov [21], Chapter 6). The most practical angle tracking signal processing techniques include phase comparison monopulse angle tracking. Monopulse provides improved antenna gain and efficiency, improved error slope performance, jamming resistance over conical scanning or sequential lobing techniques, wideband performance, and long-range performance characteristics.

The phase comparison monopulse mechanization includes a two-plane phased array antenna divided into four quadrants, constructed on a thin flat plate, and follows the principles of the interferometer and shown in Figure 6.12. There are $\lambda/2$ slots cut into a ground plane where λ is the designed operating wavelength and represented in the A-plane as a set of dashed lines.

The electromagnetic wavefront impedes the slots cut into a dielectric backing creating a voltage. The operating principle for phase comparison in

Phase B: Preliminary Design

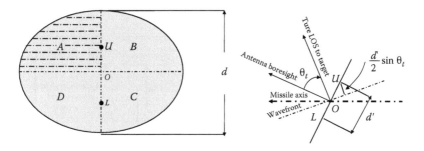

FIGURE 6.12
Phase comparison monopulse antenna representation. (Modified from James, D.A., *Radar Homing Guidance for Tactical Missiles*, Macmillan Education, Basingstoke, UK, 1986, Figure 4.6, p. 43 [20].)

two planes and using four quadrants is covered completely in the literature [20–23]. The phase difference, $\Delta\phi$, between the four elements is used to determine the angular location of the target in the following way. Angular pitch error (Δ_{el}) is defined by subtracting the addition of the upper quadrant ($A + B$) voltages from the addition of the lower quadrant ($D + C$) voltages. Angular azimuth error (Δ_{az}) is found by subtracting the left-hand pairs ($A + D$) from the right-hand pairs ($B + C$). The summation of all quadrants is given in ($A + B$) + ($C + D$). See the following equations:

$$\Delta_{el}(\theta_t) = (A + B) - (D + C) \tag{6.2}$$

$$\Delta_{az}(\theta_t) = (A + D) - (B + C) \tag{6.3}$$

$$\Sigma = (A + B) + (C + D) \tag{6.4}$$

The error voltages are used to drive the antenna servos. The LOS rate is sometimes measured by placing two orthogonally mounted rate gyroscopes to the antenna gimbals.

Tracking the target in angle then relies on computing the ratio $\delta = \Delta/\Sigma$. When $\delta = 1$, the antenna points at the target within a 3 dB beamwidth.

Figure 6.13 represents the general case for a monopulse seeker angle tracking. This block diagram includes body rate stabilization, torque motor and seeker gimbal dynamics, error signal filtering, and nonlinear limiting.

The resultant angular error rates are fed to drive torque motors that after body stabilization scan the seeker antenna toward the target and passed to the guidance computer where guidance steering commands are generated.

Active radar angular noise sources that need to be considered include radome boresight error, monopulse tracking error, glint, receiver noise, clutter, and multipath. These noise sources, detection loss sources, and jamming are presented later in this chapter when GNC is addressed.

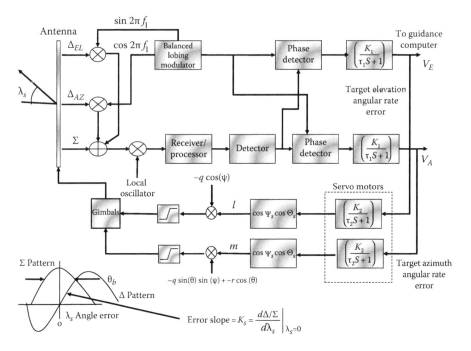

FIGURE 6.13
Monopulse seeker angle tracking block diagram.

6.5.1.2 AR Seeker Preliminary Design

The seeker design trade study here will focus on an active radar (AR)-only strategy to demonstrate the design trade process. In choosing an AR design, higher frequencies will, in general, result in shorter homing times and increased power requirements but will also provide higher resolution for tracking and the ability to use wider bandwidths for jamming resistance and discrimination. Moreover, higher frequencies with increased angular accuracies will allow smaller, lighter kill mechanisms and thus smaller lighter missiles. AR homing will also allow for an increase in firepower by reducing the AMD system resource commitments during the engagements. The somewhat recent development of cheaper higher-grade inertial reference systems and satellite-aided navigation may reduce the technical risk associated with achieving stressful handover accuracies. In short, a high-frequency AR-only seeker would be one practical AMD interceptor design strategy choice. This choice will, however, drive requirements back onto the interceptor responsiveness and handover requirements.

Assuming an AR solution is chosen, the most important design considerations are the frequency and antenna specifications. Seeker frequency will influence the remainder of the design choices throughout the seeker and should be chosen in parallel with the antenna type and design. Waveform

Phase B: Preliminary Design

selection and signal processing approaches are two designs that will follow frequency and antenna selection.

Before the seeker frequency can be chosen, the intended mission area from Figure 6.11 should be revisited. A multiple pulse repetition frequency (PRF), Ka-band AR seeker satisfies the clutter discrimination requirements for low-altitude cruise missile defense, airborne target detection capability with high-speed closing velocity estimation capability, high-angle resolution tracking capability, multiple target discrimination capability for ballistic missile defense, and complex air defense environments and may allow a target imaging option for aim point selection. Therefore, it is assumed that the trade-space study results for the illustrative air and missile defense application at hand settled on a Ka-band (~35 GHz), multiple PRF, pulse Doppler (PD), and active radar (AR) seeker design as an appropriate choice to blend performance requirements in a general air and missile defense application. See James [20, chapters 8,9].

Moreover, AMD scenarios will typically involve head-on or close to head-on engagement geometries, and air targets will likely be engaged with trajectory shaping that forces lookdown geometries. These engagement scenario constraints are the strengths of PD. The multiple PRF modes are nominally referred to as low, medium, and high PRF modes. The low PRF mode permits unambiguous range measurement and will also be unambiguous in velocity against low-speed targets possibly found in the air target environment. Although easily made stable when compared to the higher PRF modes, it is limited to shorter ranges, and other than slow targets, it is ambiguous in velocity. The medium PRF mode is exactly what is intuitively obvious to the engineer. It provides unambiguous intermediate ranges and velocities. It provides modest difficulty in retaining stability requirements, but it does not perform any of the AMD jobs very well. Were AR seeker PRF selection a political contest, medium PRF would surely be the choice of the undecided voter. Although medium PRF does not provide the specific performance desired in either range or velocity, it does provide a good transition to when either the targets are not behaving as expected or the geometries are not as favorable as the designer would like. In the end medium, PRF may be more useful in deciding what you do not want the seeker to track and therein lays the benefit of medium PRF, a potential added discrimination capability. A high PRF seeker provides unambiguous velocity measurement, and superior clutter filtering. The costs of these desirable performance features are more complex signal processing and the need to purchase high-stability components. High PRF PD is range ambiguous approaching that of continuous wave seekers as the duty factor (ratio of the transmit pulse width to pulse repetition interval) approaches one.

The PD AR seeker design trade space beyond the specific parameters already chosen (signal RF, and using a multiple PRF design) includes antenna aperture, signal bandwidth, the specific PRFs and their grouping, average power, coherent processing interval (CPI), and the thermal noise density. This design trade space will limit the ultimate performance of the seeker to measure (resolution) and track range, velocity, and angle. Details associated with

RF seeker design can be found in the literature (e.g., James [20], Maksimov and Gorgonov [21], Edde [22], Barton [23], Weidler [24], Hendeby [25], The George Washington University [26], Nathanson and Jones [27], Schleher [28], Miwa et al. [29], Mitchell and Walker [30], and Shnidman [31]). Trapp [39] provides a comprehensive PD radar analysis and missile seeker design example providing an excellent source for more in-depth AR seeker requirement analysis. Table 6.1 is reprinted from Trapp [39] to provide a readily available summary of the AR seeker design trades that need to be examined. These design trades will be addressed in some detail throughout the rest of Section 6.5.1.

Column 1 in Table 6.1 identifies important performance criteria, which should be addressed in preliminary design, and the remaining columns depict with an x the design parameters that are related to establishing the specific performance criteria. The signal-to-noise ratio is directly related to range performance criteria that can be mapped to the target mission requirements provided in Figure 6.11a through d that relates target velocity to the required seeker detection range as a function of homing time. It is important to note that the detection range calculation is actually a statistically varying process where the minimum required signal-to-noise ratio $(S:N_{min})$ to achieve a specific detection range is a function of probability of detection (P_d) and probability of false alarm (P_{fa}) as well as the target signature fluctuation properties (see Shnidman [31], Sandhu [32], and Huynen et al. [33].)

Target mission and characteristics are not directly related to $(S:N_{min})$ but are, through Figures 6.11, related to detection range. The radar range

TABLE 6.1

Active Radar Pulse Doppler Seeker System Trade Space

Affected Performance Criteria	RF	Antenna Aperture	Fundamental Radar Characteristics				Thermal Noise Density
			Signal Bandwidth	CPI	PRF	Avg. Power	
Signal-to-noise ratio	×	×		×		×	×
Range resolution			×				
Range measurement accuracy	×	×	×			×	×
Range ambiguities					×		
Velocity resolution	×				×		
Velocity measurement accuracy	×	×		×		×	×
Velocity ambiguities	×				×		
Angle resolution	×	×					
Angle measurement accuracy	×	×		×		×	×

Source: Trapp, R.L., Pulse Doppler Radar Characteristics, Limitations and Trends, FS-84-167, The JHU/APL, Howard County, MD, October 1984 [39].

equation can be expressed in many forms and is the best starting place for RF seeker preliminary design. Equation 6.5 presents a form [39] suitable for frequency (wavelength) selection and antenna, coherent processing interval (CPI), and average power design:

$$R_{max} = \left[\frac{P_l \cdot G_d^2 \cdot \sigma \cdot \lambda^2}{(4\pi)^3 \cdot KT_0 \cdot NF \cdot W_D \cdot (S/N)_{min} \cdot L} \right]^{1/4} \quad (6.5)$$

The preliminary design process begins with specifying target signature characteristics, and it is necessary to develop an assessment of the target and signal environment. Active radar frequency and polarization leading to aspect-dependent target RF signature magnitude and fluctuation characteristics (glint) are specified as a set of requirements. Target flight characteristics and the time–space correlated RF signature characteristics are further expanded as part of the requirements set. Next, the signal propagation environment is specified. Requirements are developed to operate in all-weather (rain, heavy rain, clear, etc.) and anomalous propagation environments to include clutter and multipath. Jamming environments are specified as part of the requirement set. Jamming can include standoff or self-screening systems or no jamming at all. Finally, the kill criterion has to be specified. A *hit-to-kill* vice specifying an acceptable miss distance criterion places demands on the RF frequency and homing time requirements.

The closing velocities depicted in Figure 6.11, ranging from Mach 4 to 12, can be expected. If V_C represents the closing velocity, VI_{los} the interceptor velocity along the line-of-sight vector, and VT_{los} the target velocity along the line of sight, then

$$V_C = VI_{los} + VT_{los}$$

Physics then tells us that the Doppler frequency shift F_D produced by V_C is shown in the following equation:

$$F_D = \frac{2 \cdot V_C}{\lambda} \quad (6.6)$$

After doing the math, the Doppler filter bandwidth, W_d, will have to adapt from 32.5 to 145 kHz assuming a seeker center frequency of 34.5 GHz with a 1 GHz bandwidth to detect and track the target family described in Figure 6.11 with some reasonable margin.

The antenna gain (transmit and receive) is written as follows:

$$G_d = \frac{4 \cdot \pi \cdot A_e}{\lambda^2}$$

The antenna diameter (d) and efficiency factor (η) are used to compute A_e, the effective aperture of the antenna:

$$A_e = \frac{\eta \cdot \pi \cdot d^2}{4}$$

The minimum discernable signal, S_{min}, represents how much power is required for detection without jamming or clutter (clear environment). The average noise power at the receiver is $N = K \cdot T_0 \cdot W_d$ and S_{min} can be written as follows:

$$S_{min} = NF \cdot (N \cdot (S{:}N)_{min})$$

The 3 dB beamwidth is a measure of angular resolution and subsequently the radar's ability to resolve multiple targets and suppress unwanted signals. According to Farrell and Taylor [40], the angle measurement accuracy is proportional to beamwidth and inversely proportional to the square root of the S:N. The 3 dB beamwidth is determined from the RF wavelength and the antenna area, written as follows:

$$\theta_b = \frac{\lambda}{\sqrt{\pi \cdot (d^2/4)}}$$

The achievable range resolution and estimation accuracy are a function of the radar's ability to measure time. Also according to Farrell and Taylor [40], the time measurement capability is inversely proportional to the signal bandwidth and the square root of S:N. This mathematically implies that the radar range estimation accuracy increases with signal bandwidth assuming a fixed S:N. Pulse width (PW) is inversely proportional to signal bandwidth and shortening PW makes a convenient way to improve radar range estimation accuracy.

The coherent processing interval (CPI) is defined as the time duration over which the radar returns are coherently integrated. CPI can be used to improve S:N but is the means to improve target velocity measurement accuracy. CPI is inversely proportional to Doppler resolution bandwidth indicating that Doppler frequency measurement accuracy (velocity measurement accuracy) is inversely proportional to CPI and the square root of S:N according to Farrell and Taylor [40]. For the purposes of preliminary design, it is convenient to assume that the Doppler bandwidth W_d is approximately equivalent to the reciprocal of the coherent processing interval (CPI).

To begin developing a preliminary design seeker solution, a requirements flow down from the TLR must be completed. The first requirement needed is the handover error volume. Handover error must be first defined in terms of an angular uncertainty volume (ψ_{az}, ψ_{el}) and range to target (R). Next, the amount of search time for acquisition (T_{sc}) is established; the target characteristics include speed (V_T), signature magnitude range (σ_{min} to σ_{max}), and signature fluctuation characteristics (the use of Swerling terminology is suggested [42,43]). Finally, the cumulative probability of detection (P_{dc}) and probability of false alarm are specified performance requirements. Additional requirements will flow down from various design budgets. Mass and size budgets will limit the seeker and antenna properties accordingly. Technology constraints will flow from a risk analysis and will limit transmitter power, receiver signal processing techniques, noise and stability characteristics, and antenna and tracking technology among other details. Additional design constraints concerning assumptions on the engagement environment will impact predicting signal and processing losses.

In the following active radar seeker preliminary design example, refer to Figure 6.11a–d. Targets will span subsonic to hypersonic ($M > 5$) velocity regime and they will occupy terrain following or sea skimming to upper endoatmospheric altitude regimes. It is assumed that target signatures for this example design will extend from as low as −20 to +5 dBsm and will exhibit Swerling 0 and 1 fluctuation losses. A single interceptor, or even seeker, design may not prove to be a reasonable top-level design approach given the extent of the target requirements. Therefore, it is prudent to begin by singling out the most stressful target requirements and then move to the environmental stressors that cause performance degradation such as anomalous propagation, jamming, and clutter. The design can then be modified to accommodate graceful performance degradation. Design iterations will eventually lead to the point where the engineering trade solutions will not close and will require a decision to choose either a different seeker strategy or an independent interceptor design or both.

The most stressful target is the fastest and lowest signature, respectively. Examining Figure 6.11's illustrative charts indicates that a Mach 7 target is the most stressful case. Assuming that in the TLR flow down the keep-out zone imposed on the AMD system is relatively large and the mission extends beyond self-defense, the fastest interceptor design option will be chosen for the first set of iterations. The first design iteration will be to resolve the lowest target RF signature that can be engaged while meeting the other design and risk constraints discussed earlier.

The first seeker design iteration will assume employing a low-risk phased array antenna, allowing a number of independent target looks or CPI detection opportunities. The illustrative AMD system will be required to provide a worst-case condition of $\Psi_{el}/\Psi_{az} = 12°$ by 12°, 1-pulse width range handover error volume at 25 km range to go from the target. If a terminal Mach 7, SRBM target, and a Mach 5 interceptor are assumed and the altitude-dependent average speed of sound is 300 m/s, then the time to go at handover is 7 seconds.

The seeker design will need to be unambiguous in velocity and have a primary high PRF mode of 921 kHz ($V_C > 3600$ m/s) and a 0.1 μs pulse width

TABLE 6.2

Pulse Doppler Active Radar Seeker Design Specifications and Performance

$A_e = 0.0245$ [eta = 0.5, $d = 0.25$ m]
$L = 10$ [10 dB] $T_0 = 290$ K
$NF = 10$ [10 dB] $k = 1.3804$E–23
$W_d = 550$ [Hz Doppler resolution] $c = 299{,}792{,}458$ m/s
$S/N_{min} = 10.5$ [10.21 dB] $kT_0 = 4.0031$E–21
$S_{min} = 2.3118$E–16 [–156.36 dB]
$\lambda = 0.00869$ m [$freq = 3.45$E+10 Hz]
$\theta_b = 0.06$ rad   3.18°   3 dB beamwidth (circular antenna)
$Gd = 4084.57$ 36.11 dB   Antenna gain

to provide a 10 MHz receiver bandwidth. The Doppler bandwidth is 550 Hz, the CPI is 2 ms, and the duty factor is 9.2%. The remaining illustrative seeker design parameters are given in Table 6.2.

The antenna diameter is assumed to be 0.25 m with an efficiency of 0.5, a gain of 36.11 dB, and a 3 dB beamwidth of 3.18°. The Stefan–Boltzmann constant is k, and assuming reference noise temperature $T_0 = 290$ K, then $kT_0 = 4 \times 10^{-21}$. It is assumed that a receiver noise figure (NF) of 10 is low risk. Losses are assumed to include RF system losses (L_{RF}), signal processing losses (L_{SP}), and beam pattern losses (L_{BP}) and together equal 10.

The minimum required signal-to-noise ratio ($S{:}N_{min}$) is found from the handover error requirement and the following analysis. The seeker design has a 3.2°, 3 dB beamwidth to cover a 12 × 12 degree uncertainty volume. Using the relationship $N_L = V/2(BW_{3\ dB})$, the uncertainty volume can be covered with 23 separate 3.2°, 3 dB beamwidth beam positions. Assuming that half of the time to go is available for acquisition and half for homing, then with a CPI of 2 ms and a search time of 3.5 seconds, the search volume can be covered with six independent looks per beam position. Imposing a requirement for a cumulative probability of detection (P_{CD}) of 0.9 at 25 km range to go, where

$$P_{CD} = 1 - (1 - P_D)^{NL}$$

six independent looks allow for a single look P_D of 0.3. Specifying a probability of false alarm (P_{fa}) of 10^{-6} and using the signal-to-noise ratio versus P_D chart from Blake [77], chapter 2, pp. 2–19, for a nonfluctuating target, the single pulse, $S{:}N_{min}$, is found to be 10.25 dB. Table 6.3 provides the acquisition performance estimates for this preliminary design example.

The parametric average transmit power, P_t, is the first row of Table 6.3 starting at 50 W and ending at 2500 W. The parametric target signature, *sigma*, begins the acquisition range performance summary and varies between 0.01 and 5 m². An analysis of the results of this table reveals that the preliminary

Phase B: Preliminary Design

TABLE 6.3
Pulse Doppler Active Radar Seeker Acquisition Performance for Table 6.2

P_t (watts)	50	100	250	500	1000	1500	2500
Acquisition Range Performance (km)							
Sigma	0.01	0.1	0.25	0.5	0.75	1	5
RNG(PI1)	3.42	6.09	7.65	9.10	10.07	10.82	16.19
RNG(PI2)	4.07	7.24	9.10	10.82	11.98	12.87	19.25
RNG(PI3)	5.12	9.10	11.45	13.61	15.06	16.19	24.21
RNG(PI4)	6.09	10.82	13.61	16.19	17.91	19.25	28.78
RNG(PI5)	7.24	12.87	16.19	19.25	21.30	22.89	34.23
RNG(PI6)	8.01	14.25	17.91	21.30	23.58	25.33	37.88
RNG(PI7)	9.10	16.19	20.35	24.21	26.79	28.78	43.04

seeker design presented would capture the proposed target with 3.5 seconds of homing time remaining when the target signature is greater than 0.5 m² (nonfluctuating) and having an average transmit power of 2.5 kW or greater. When target signatures are greater than 1 m², a P_t of 1.5 kW or greater will meet the design requirements. However, if the target set is expected to be smaller than 0.5 m², the P_t required is greater than 2.5 kW shown on this graphic. One design alternative would be to trade lower frequency (e.g., Ku band) for the power requirement. However, the lower frequency will increase the 3 dB beamwidth making the system more vulnerable to jamming and also less accurate adversely impacting the kill strategy and impacting other interceptor design choices.

When dealing with atmospheric and especially low-altitude intercepts, a number of environmental problems including adverse weather, clutter, and multipath will ultimately limit this performance. The following paragraphs will discuss signal transmission losses and jamming. The remaining noise sources important to consider when designing a seeker integrated with a guidance system will be handled in the guidance, navigation, and control section.

6.5.1.3 Signal Transmission Losses

There are many sources of RF noise in the environment. Target fluctuation is a major source of transmission losses. Complex target shapes, corners, and appendages result in a large number of aspect- and polarization-dependent primary RF scattering centers that can represent the reflected signal radar cross section (RCS) transmission. Target motion–induced aspect angle changes relative to a tracking seeker will induce randomly varying RCS with time. Marcum–Swerling (MS) models (see, e.g., Schleher [28], Mitchell and Walker [30], Marcum and Swerling [42], and Swerling [43]) have often been used to statistically describe the RCS fluctuation characteristics of a target when performing radar range performance calculations. MS models can be divided into five categories and refer to a specific signal processing model that can be found in the literature, and the target RCS can be described by a chi-squared distribution. MS-0 refers to a nonfluctuating target; MS-1 and MS-3 refer to a scan-to-scan or slowly fluctuating target; MS-2 and MS-4 refer to a pulse-to-pulse or rapidly fluctuating target. MS-1 and MS-2 are called a Rayleigh target (see Skolnik [41], pp. 2–18) while MS-3 and MS-4 are non-Rayleigh targets better represented by log-normal distributions (see Skolnik [41], pp. 2–19). The slowly fluctuating target assumes that the time-dependent RCS values are statistically independent on a scan-to-scan basis but are constant on a pulse-to-pulse basis. A rapidly fluctuating target RCS is statistically independent on a pulse-to-pulse basis within one 3 dB beamwidth during one CPI.

The AR design can be improved to handle various target fluctuating characteristics by including uncorrelated multiple independent CPI period looks. Each uncorrelated look occurs when multiple CPI sets of pulses are noncoherently integrated or averaged. The AR seeker designer will choose the target dwell time, CPI, and the number of uncorrelated independent looks to integrate. The designer will choose, among other things, whether to have an electronically or mechanically scanned phased array antenna. Each option comes with advantages and disadvantages having an impact on other design choices that will affect the overall interceptor mission capability and limitations.

Atmospheric attenuation is another form of signal transmission loss and is typically measured in dB/km. Atmospheric conditions will vary specifically and generally with geographical location and altitude but never with absolute certainty. Subsequently, frequency, power, and polarization design trades will need to be considered as a function of mission and operating area. Generally, signal transmission loss increases proportionally with frequency but is not monotonic. For AR purposes, the atmosphere can be described according to humidity (atmospheric water content as a percentage), oxygen content, and precipitation. Relative attenuation nulls occur at high frequencies near 30–40 GHz and near 90–95 GHz, while spikes occur between 20–25 GHz and 50–70 GHz. Precipitation can and will vary as a function of altitude, and for ballistic missile defense application, this can be a design problem that needs to be assessed. For example, rainfall rates generally decrease as altitude increases. Precipitation causes the greatest amount of atmospheric attenuation, and the greater the precipitation rate, the greater the attenuation problem. Charts characterizing frequency versus attenuation for various atmospheric conditions can be found in the literature (e.g., Skolnik [41], pp. 2-51–2-59]).

6.5.1.4 Jamming

RF jamming may always be a potential technique used to deny range and/or angle detection and tracking. Jamming signals can be divided into

Phase B: Preliminary Design

two categories: denial and deception. Deception jamming may try to mimic tracking signals using coherent, digital RF memory (DRFM) techniques to replicate or repeat return echoes from offensive system in order to fool the interceptor range and/or angle tracker into tracking a nonexisting object and thereby lose track of the real target. Denial techniques are used to hide the radar echoes off of the offensive system by saturating the radar receiver with dense or barrage noise in the seeker frequency band and over the entire receiver bandwidth. One way to reduce the effectiveness of the jammer in this case is to employ a wideband (>1 GHz) seeker system with pulse-to-pulse frequency agility [23]. This analysis can be quantified by employing the equations on jamming proposed by Schleher [28]. Figure 6.14 shows the burnthrough range performance of a 35 GHz active radar seeker when a jammer of specified power and 500 MHz bandwidth attempts to deny its range as a function of jammer carrier (target) radar cross section. Figure 6.15 then shows the same graphic relationship where a 1 GHz bandwidth jammer is required to cover the seeker bandwidth. It is apparent from the graphics that spreading the jammer across a larger bandwidth reduces its effectiveness. Therefore, the receiver bandwidth needs to be considered in any engagement environment where jamming is likely or even possible to occur.

6.5.2 Translational and Attitude Response Preliminary Design

The translational and attitude response preliminary design development follows from the engagement boundary envelope requirements, target specifications, and the seeker preliminary design. Moreover, the translational

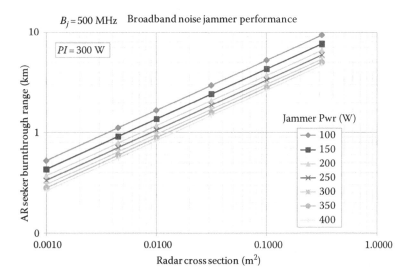

FIGURE 6.14
Narrowband MMW AR seeker versus jammer performance.

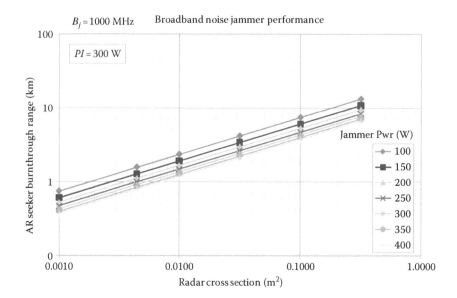

FIGURE 6.15
Wideband MMW AR seeker versus jammer performance.

response requirements will depend on an efficient midcourse guidance strategy design and handover requirements discussed in Section 6.4 (part of the seeker requirements flow down) and will subsequently influence attitude response requirements. The proposed translational and attitude response preliminary design process is provided in Figure 6.16 and is discussed in the following paragraphs.

Target specifications drive translational and attitude response design in two ways and impose requirements in both the *reach* and *end game*. First, target speed, maneuverability, and agility will place energy requirements on the interceptor during homing or end game. The P_{ssk} requirement is directly influenced by the interceptor homing time constant to target maneuver time constant ratio. As the interceptor velocity increases (translational requirement), so does this ratio during end game. A 3:1 ratio is desirable. Second, the target signature directly influences homing time. The seeker design determines how much homing time is available and homing time is inversely related to miss distance and thus directly with P_{ssk}. The length of available homing time is directly proportional to the required missile homing time constant. Therefore, if effort and cost were invested in the seeker design to ensure long homing times, then the missile homing time constant requirement is somewhat relaxed. And the opposite is true if only a minimal amount of homing time is designed to be available. The trade where performance and cost make the most sense will be determined on a program-by-program basis. This discussion should make it obvious why seeker preliminary

Phase B: Preliminary Design

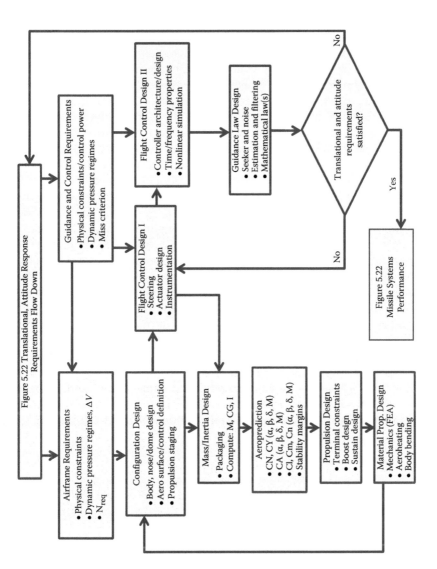

FIGURE 6.16
Translational and attitude preliminary design process.

design should be accomplished first. Iteration will be required to settle on a satisfactory preliminary design.

6.5.3 Airframe Requirements

The airframe preliminary design will commence once the Figure 6.16 requirements flow-down process completes its first iteration. The CDS interface requirement will include a launcher mechanism concept already determined. The launcher concept will dictate volume, length, diameter, and weight constraints. Other constraints from policy considerations possibly limiting the size, speed, and range will also flow down from Figure 5.22. The Mach–altitude boundary, Figure 5.21, must be combined with the altitude–range boundary, Figure 5.20, to establish intercept Mach requirements that in turn establish the flight regimes of interest that include dynamic pressure, Reynolds number, and Mach number. The dynamic pressure ($Q = \rho \cdot V^2/2$) regime is a major concern as it defines aerodynamic forces and moments as a function of configuration. Reynolds number, $R_e = \rho \cdot V \cdot \ell / \mu$, defines the nature of the aerodynamic boundary layer viscous flow. Specifically, critical R_e defines the transition point from laminar to turbulent flow by identifying the point in flight where a significant increase in drag and body temperature exists. Mach number is defined as the ratio between the total velocity vector magnitude and the local speed of sound (a), where $a = \sqrt{\gamma \cdot R \cdot T}$. Interceptor missile flight will almost certainly be bounded, after initial boost conditions, in the high supersonic to hypersonic flight (see Table 6.4) with the associated Mach regimes.

The next requirement to derive is the airframe normal force, N_{req}, in terms of the flight regime and the potential target maneuverability performance at the outer edges of the engagement envelopes. A reasonable rule of thumb is that the endoatmospheric interceptor missile must have a three-to-one ratio of maneuverability advantage over the target at end game. The exoatmospheric end game requires more of an energy management strategy. The aerodynamic maneuver requirement is given in Equation 6.1 and is specified in units of "g's":

$$C_{Nreq} = \frac{N_{req} \cdot W}{S \cdot Q} \tag{6.7}$$

TABLE 6.4

Interceptor Missile Flight Regime Requirements

Mach Regime	Requirement	M, Low End	M, High End
Subsonic	Boost	0	1.0
Transonic	Boost	0.85	1.2
Low supersonic	Boost	1.2	2.0
High supersonic	Boost/sustain	2.0	<5.0
Hypersonic	Sustain	5	5+

Q is defined by the engagement boundaries, while weight, W, and aerodynamic reference area S ($S = \pi \cdot d^2/4$) are the trade space. The engagement boundaries will ultimately be in play as a potential trade space.

6.5.4 Configuration Design

The purpose of the configuration design process in the first iteration is to develop a preliminary configuration that is likely to meet all of the constraints, addressing the functional, performance, and interface requirements that have flowed down. The preliminary configuration is based primarily on the requirements flow-down process and some preliminary aerodynamic predictions. There are three primary missile body sections—forebody, midbody, and aft body—that need to be defined. Drag is a primary configuration design driver. Drag has three components—pressure, friction, and base—that need to be managed. The configuration is the primary drag management tool. The forebody or nose will be some form of dome used to cover the sensor during atmospheric flight and, besides being either electrical or optically conductive, must provide adequate aerodynamic, thermodynamic, and volumetric properties. The nose shape is likely to be driven by the sensor employed but must be a compromise design that minimizes drag, tolerates high heating, and provides the necessary lift characteristics. Improved drag characteristics are achieved with high fineness ratios (length to diameter), but in general an ogive nose configuration is the best compromise design, offering a greater volume for packaging, structural superiority, and adequate drag characteristics. A tangent-ogive dome will likely provide the best combinations for achieving all of these properties in the flight regimes of interest [7].

Thin, slender body shapes are preferred over short stout ones, and aerodynamic surfaces should have sharp leading and trailing edges. The midbody configuration will be body–tail (BT), body–wing–tail (BWT), body–canard–wing (BCW), or some other combination of body, lifting, and control devices. Wings are part of the trade space. There are advantages and disadvantages to having a winged design. Wings add not only weight but also structural integrity. Structural integrity decreases body flexure that increases the complexity and weight of the control system. The mission design and control strategy will likely drive the inclusion of wings or not. The aft body should have a tapered boat–tail to minimize base drag during power-off flight by reducing the base area of the missile. Some of the trades influencing boat–tail design include increased aft-end lift with increased tapper causing a destabilizing effect requiring increased control surface and an increase in drag.

A steering policy must be determined early in the configuration design process. A skid-to-turn (STT) versus preferred-orientation-control (POC) policy is the trade space. POC requires a bank-to-turn steering while the STT policy is either a roll rate control system or a roll attitude control system. Control implementation is the next design decision. Control or steering is accomplished through either aerodynamic or propulsive forces.

Aerodynamic steering involves tail, wing, or canard controls. Tail control has generally been preferred for aerodynamic systems when the trades of control authority, drag, packaging, and responsiveness are concerned. Tail steering is not the best approach in any one of these areas but is superior in the aggregate. Propulsive control involves reaction control systems (RCSs) or thrust vector control (TVC) systems. Hemsch and Nielsen [15] provides the challenges, dynamics, and flight control of employing RCS, TVC, or in combination. Blended aerodynamic and propulsive control systems have been employed in the PAC-3 system to take advantage of the superior performance of aerodynamic control when in lower atmospheres and the advantages of RCS in end game and less dense atmospheres where responsiveness is critical. Other missiles use TVC for boost to quickly align the velocity vector and then transition to aerodynamic control.

The approximate analytical expression given in the following equation addressed by Chin [7], Nesline and Nesline [10], and Moore [11] allows the configuration process to begin:

$$C_N = 2 \cdot \alpha + C_{Dc} \cdot \left(\frac{S_P}{S}\right) \cdot \alpha^2 + 8 \cdot \left[\frac{S_t}{\left[S \cdot \sqrt{M^2 - 1}\right]}\right] \cdot (\alpha + \delta) + 8 \cdot \left[\frac{S_W}{\left[S \cdot \sqrt{M^2 - 1}\right]}\right] \cdot \alpha$$

(6.8)

To consider RCS and/or TVC, consult Hemsch and Nielsen [15] for the terms that can be added to Equation 6.8. Equation 6.8 assumes a BWT configuration, although a canard control could be used in place of the tail component. If a body–tail configuration is desired, the wing term would simply be set to zero.

Figure 6.17 presents a notional baseline body–wing–tail configuration intended for endoatmospheric intercept missions. An interceptor missile configuration design intended for exoatmospheric missile defense will require a

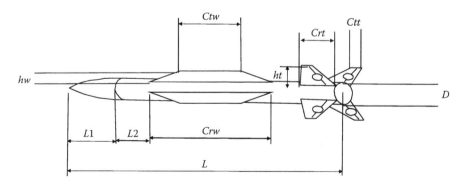

FIGURE 6.17
Notional missile configuration concept.

Phase B: Preliminary Design

different vehicle design definition than will a low-altitude air defense interceptor missile, for example.

It is necessary at this point to discuss that interceptor missile airframe time constant or attitude response requirement has a dominate effect in achieving desirable miss distances (leading to P_{ssk} for a given kill strategy) and is the single most important end-game parameter in hit-to-kill strategies. In the proposed process, the airframe time constant will develop as the preliminary design iterations mature and engagement simulation results are achieved. The definition for the parameters shown in Equation 6.8 and Figure 6.17 and terms for conducting a preliminary aerodynamic configuration trim and time constant analysis are provided in Table 6.5.

Details explaining this set of equations and terms can be found in [3,7,11]. It is important to note that the analysis given in Table 6.5 cannot be accomplished without including a total mass and inertia budget as part of the requirements flow-down process along with the external configuration parameters indicated. Producing a mass and inertia estimate based on a subsystem breakdown is covered in Section 6.4.3. Table 6.6 provides example configuration parameters associated with Figure 6.10 and the associated aerodynamic trim results are shown in Figure 6.18. The example configuration

TABLE 6.5

Aerodynamic Model to Estimate Preliminary Configuration Properties

Aerodynamic Model to Estimate Acceleration Limit			
$N = CN * S * Q/W$	$CN = 2\alpha + C_{Dc} * Sp/S\alpha^2 + [8 * S_{tail} (\alpha + \delta)/S] * \text{sqrt}(M^2 - 1) + [8 * S_{wing}/S] * [\alpha/\text{sqrt}(M^2 - 1)]$		
Sp = Missile platform area	$Sp = (L - L1) * D + 2/3 * L1 * D$	S = Reference area	$S = \Pi d^2/4$
C_{Dc} = Cross-flow coefficient	See Moore [11]	AFTC: $t_\alpha = M_\delta/(M_\alpha * Z_\delta - Z_\alpha * M_\delta)$	Missile airframe time constant
S_{tail} = Tail area	$S_{tail} = 1/2 * ht$ $(Crt + Ctt)$	$Z_\alpha = -g * Q * S * CN_\alpha/W * Vm$	$M_\alpha = -g * Q * S * d * Cm_\alpha/Iyy$
S_{wing} = Wing area	$S_{wing} = 1/2 * hw * (Crw + Ctw)$	$Z_\delta = -g * Q * S * CN_\delta/W * Vm$	$M_\delta = -g * Q * S * d * Cm_\delta/Iyy$
Aerodynamic Trim Analysis			
$CM = 2 * a[XCG - XCPN/D] + 1.5 * Sp * a2/S * [XCG - XCPB/D] + 8 * a * S_{wing}/[(\text{sqrt}(M^2 - 1)) * S] * [XCG - XCPW/D] + 8 * (\alpha + \delta) * S_{tail}/[\text{sqrt} (M^2 - 1) * S * D] * [XCG - XHL/D]$			
$XCG = L/2$			
$XCPN = (2/3) * L1$			
$XCPB = L1 + L/2$			
$XCPW = L1 + L2 + 0.7 * Crw + 0.2 * Ctw$	XCP wing is assumed to be the *cg* of the missile.		
$XHL = L - 0.3Crt - 0.2 * Ctt$			
XCP tail $= XHL$ (assumption)			

TABLE 6.6

Preliminary Configuration Parameters

Parameter	Value
Length (m)	4.80
Diameter (m)	0.35
L1 (m)	0.70
L2 (m)	1.60
Crw (m)	2.30
Ctw (m)	1.80
Crt (m)	0.37
Ctt (m)	0.10
hw (m)	0.15
ht (m)	0.50
Weight (N)	7000.00
Iyy (kg-m²)—burnout	900.00
C_{Dc}	1.50

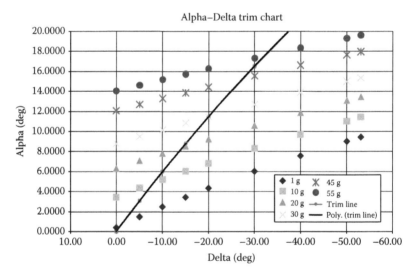

FIGURE 6.18
Aerodynamic trim results.

under study produces about 30 g's trim acceleration with 10° angle of attack (AOA) and 18° control surface deflection at sea-level conditions and Mach 2.1. Results beyond 10° AOA should be ignored when using analytical approximations such as Equation 6.8.

Air and missile defense missions demand not only highly maneuverable intercept missiles but also rapidly responding airframes sometimes referred

Phase B: Preliminary Design

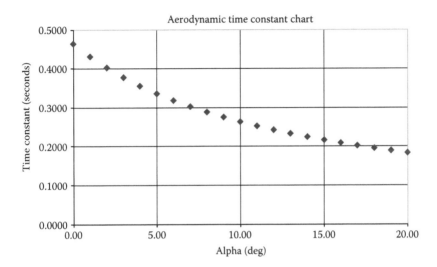

FIGURE 6.19
Representative airframe time constant.

to as jerk requirement. The intercept missile must be able to rapidly develop a high load factor. Load factor is the amount of available lift to weight ratio. The jerk requirement is a function of configuration details (BT, BWT, etc.), stability margin, and control power. The aerodynamic or airframe time constant (τ_α) is the performance parameter used to measure the design requirement. τ_α is defined mathematically in Table 6.5 and is a measure of the amount of time it takes to turn the missile velocity vector through an equivalent AOA. Figure 6.19 presents a representative τ_α as a function of AOA at constant sea-level, Mach 2.1 for the Figure 6.17 configuration using the approximations given in Table 6.5.

For this example, τ_α varies from 0.47 to 0.25 seconds between 0° and 10° AOA. Airframe responsiveness, τ_α, is only one component of the overall missile time constant τ. According to Equation 6.9 [13], τ is the measure of the interceptor missile's ability to respond to guidance errors:

$$\tau = \tau_{FCS} + \tau_S + \tau_n + N\left(\frac{V_c}{V_m}\right)\tau_\alpha R \qquad (6.9)$$

The missile time constant, τ, is shown to be a linear combination of vehicle stability and control and body-bending frequency time constant (τ_{FCS}), seeker tracking loop time constant (τ_S), and the product of the effective guidance navigation ratio (N), the ratio between closing velocity and the interceptor missile velocity, V_c/V_m, radome boresight error slope (R), and τ_α. The airframe component will be the slowest component having the most significant impact on the overall time constant and ultimately miss distance.

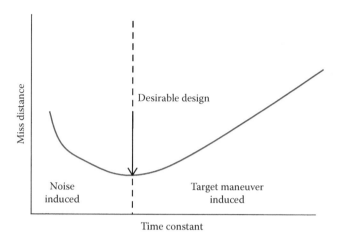

FIGURE 6.20
Miss distance relationship to time constant.

Miss distance is the interceptor missile's overall performance requirement that missile time constant influences. Figure 6.20 shows a representation of the relationship between time constant and miss distance.

Noise-induced miss distance results if the time constant is too small and target maneuver–induced miss distance results if the time constant is too large. Design requires a balanced integrated missile systems approach. The airframe design is typically made to be as responsive as possible and the remaining terms are used to tune the system. Constraints, other competing requirements, and technology risks may preclude reaching the desired time constant design.

6.5.5 Mass and Inertia Design

Once the vehicle configuration design concept is formulated, initial subsystem mass estimates are developed based on an overall mass budget requirements flow down from the TLR, engagement boundary conditions, and vehicle sizing constraints. Various approaches exist to develop these preliminary design requirements. Fleeman [5] and Chin [7], respectively, provide additional and detailed approaches to produce vehicle sizing, mass, and inertia estimates.

Figure 6.21 presents a notional interceptor missile subsystem packaging approach. Length stations are provided in percentage of overall length.

Table 6.7 presents a preliminary interceptor missile weight, balance, and inertia budget template to begin an analysis. All quantities are provided in budget percentages. A total weight budget, center-of-gravity (CG) budget, and a moment-of-inertia (MOI) budget and its components are part of the requirements flow-down process. The term, dy^2, is a squared measure of

Phase B: Preliminary Design

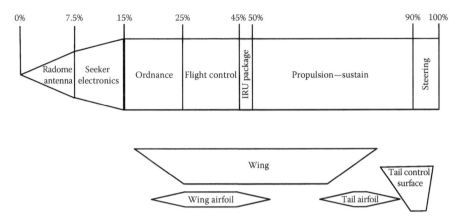

FIGURE 6.21
Notional interceptor subsystem packaging.

length between the subsystem CG and the full-up round CG. The parallel axis theorem is then used to compute the MOI budget component as indicated by the equation in column 6.

There are six primary subsystems identified for this example system. The example interceptor missile in Table 6.7 has a boost and a sustain propulsion system where the boost system separates some short time after launch once a required ΔV is obtained. The numbers in the weight column are shown as possible percentages for the subsystem components and assume a total weight or mass budget is part of a requirements flow-down process. The numbers shown in the CG column are percentages of a total length. The template is not definitive but may offer a baseline condition to begin analysis.

6.5.6 Aeroprediction

Aerodynamic requirements for an interceptor missile are flow-down range, maneuverability, agility, and velocity requirements. Six-degree-of-freedom, three force (C_X, C_Z, and C_Y) and three moment (C_l, C_m, and C_n), coefficient predictions are required to complete this requirement study. The purpose of developing these predictions is to complete the equations of motion that are used in the flight simulation and to design the flight control system. The missile body coordinate frame is employed for the coefficients. Chapter 8 provides some of the mathematical detail required to understand these coefficients and how they relate to the missile configuration. Krieger and Williams [13] provides important information and references concerning functional relationships between aerodynamic phenomena and prediction requirements. Nielsen [14] provides complete treatment of these coefficients and their relationship to the equations of motion and to FCS design.

TABLE 6.7

Interceptor Missile Weight, Balance, and Inertia Budget

Component	Weight (% Total)	X_{cg} wrt Nose Tip (% Body)	dy^2	Iyy_0	$Iyy = Iyy_0 + md^2$
Nose	1.00	7%	26%	1%	5%
Radome	0.50				
Antenna and mechanisms	0.50				
Guidance	5.00	12%	21%	1%	15%
Seeker electronics	2.00				
Power supply	0.50				
Structure	0.50				
Ordnance	11.00	20%	15%	1%	15%
Warhead	6.00				
Safe and arming device	1.00				
Fuzing package	3.00				
Flight control	8.00	40%	3%	1%	10%
Control computer	2.00				
Battery/APU	3.00				
IRU	0.40				
Structure	0.60				
Propulsion—sustain	60.00	80%	5%	90%	48%
Solid propellant	35.00	*Burnout*	0.00		30%
Rocket motor case	20.00				
Inert material	2.00				
Nozzle	3.00				
Steering and aero devices	15.00			6%	7%
Dorsal fins (4)	8.50	55%	0%		
Control surface fins (4) + actuators	1.50	90%	10%		
Ignition	100.00	58%			100%
Burnout	63.00	45%			
Propulsion boost	30.00				
Solid propellant	20.00				
Rocket motor case	5.00				
Inert material	2.00				
Nozzle	3.00				

Preliminary aerodynamic estimates may be obtained from Missile DATCOM [8] and AP09 [9]. Both of these codes are widely used in the profession and include empirical, semiempirical, and theoretical techniques. Both codes will usually provide acceptable force and moment data leading up to preliminary design review. The proper approach when using either of these codes is to first obtain verified wind-tunnel data or full Navier–Stokes CFD data against similar configurations of interest and use these data sets for calibration. Before employing either or both codes, reasonable matches should be obtained against a set of similar configurations under similar flight regimes of interest. This will help ensure you are capturing the configurations properly and have settled on appropriate settings. For example, settings such as at what Mach number to employ, the DATCOM second-order shock expansion techniques will vary depending on configuration details and other parameters. Specifically, the TriService/NASA database, discussed by Krieger and Williams [13], can be an initial source of archived aerodynamic data. Moreover, it is recommended that Missile DATCOM and AP09 and/or other similar codes be employed simultaneously for comparison purposes.

6.5.7 Propulsion Design

A propulsion design would be developed based on the velocity change (delta-V or ΔV) requirement derived from the mission and target requirements. Earlier, it was mentioned that the timeline is critical to achieving air and missile defense objectives and having sufficient energy at end game to achieve adequate miss distance against the target set. Therefore, it is not sufficient to design to average ΔV alone but terminal velocity on target should be considered as well. This combined requirement set will influence the designed velocity magnitude and time profile. This design will also be iterative. The propulsion design trade space will include whether a solid or air-breathing system is required and whether multiple stages are required. Assuming a solid rocket motor stack-up or an air breathing jet engine (probably a ramjet) is assessed to be sufficient to meet the design performance requirements, either Fleeman [5], Sutton and Biblarz [17] or Crassidis and Junkins [64] will be appropriate to develop a preliminary design.

Although improved performance can be obtained from air-breathing propulsion systems [5], their complexity, expense, and the packaging issues usually prohibit them from being a design option for AMD missions. Solid rocket motors (SRMs) operate over any Mach number and are insensitive to altitude and angle of attack. These three issues usually make the SRM the optimum design choice for interceptor missiles. For example, angle-of-attack sensitivity, a critical performance consideration, will limit end-game maneuverability. Weight is always a concern since SRM propellant is heavier than air-breathing fuel. By using staging techniques, the SRM is typically lighter than an air-breathing configuration once the propellant is expended. Therefore, the remaining treatment in this book will focus on the SRM.

Time-dependent thrust is the primary performance measure for the propulsion system. For the SRM, thrust can be related to I_{sp} [5] where

$$Thrust = I_{sp} \cdot \dot{w}_f \qquad (6.10)$$

I_{sp} is defined as the amount of thrust produced per unit weight of propellant expended. The typical values of I_{sp} in units of seconds range between 200 and 350. Theoretical values for solid propellants are limited near 500 seconds [17].

The next most important performance parameter is velocity gain or achieved also referred to as ΔV. Velocity gain can be related to I_{sp} by the relationship known as the ideal velocity gain equation or the *rocket equation* and given in modified form in the following equation:

$$\Delta V = I_{sp} \cdot g \cdot \ln(MR) \cdot K_p - t \cdot g \cdot \sin(\gamma) \tag{6.11}$$

The term g is the acceleration due to gravity, MR refers to the mass ratio, and K_p ($0 < K_p < 1$) is a performance constant that modifies the ideal velocity gain equation to account for drag. The term γ is flight path angle and the right-hand term accounts for deceleration due to gravity. Mass ratio can be obtained as a function of I_{sp} for horizontal flight ($\gamma = 0$) with a pure boost, sea level, velocity gain of Mach 3.5, and then coast, consistent with a short-range intercept missile. Figure 6.22 shows a plot of mass ratio for three assumed K_p drag modifiers, 0.7, 0.8, and 0.9.

A high drag missile would be represented well by $K_p = 0.7$ and refers to a velocity loss of 30% from an ideal gain case. Assuming a moderately clean missile configuration and a propellant with an I_{sp} of 250 seconds, shown by the circle in Figure 6.22, a mass ratio of 1.85 would be required to meet these performance specifications. Mass ratio is defined mathematically:

$$MR = \frac{M_{ignition}}{M_{ignition} - M_{propellant}} \tag{6.12}$$

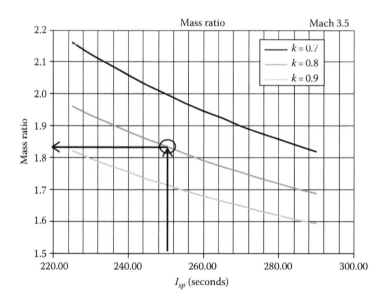

FIGURE 6.22
Mass ratio calculation for specified performance.

TABLE 6.8
Propulsion Design Parameters

Motor Case Design	Length (m) Diameter (m) Volume (cm³)	2.0 0.4 240,528
Propellant Type	I_{sp} (seconds)	Density (kg/cm³)
DB	220–230	0.001605
DB/AP/AL	260–265	0.001785
DB/AP-HMX/AL	265–270	0.001785

Table 6.8 presents an example set of motor case dimensions and propellant characteristics from Sutton and Biblarz [17]. A required propellant mass of 360 kg is obtained when assuming a propellant loading of 80% as shown in Figure 6.23 if either of the higher-performing propellants is also assumed.

Using the required mass ratio, the propellant loading analysis, and Equation 6.12, the ignition and burnout masses are presented in Figure 6.24. The vehicle ignition mass is 800 kg, and the vehicle burnout mass is 440 kg for this example.

Burn time and vacuum thrust can next be determined to provide the baseline propulsion preliminary design example. Figure 6.25 provides a combined analysis where the designer would resolve a burn time from propellant characteristics and volume loading and then resolve thrust from the other dependencies. In this example, a burn time of 30 seconds is calculated to be reasonable, and with the assumed propellant 30,000 N of average, vacuum thrust is required to produce the desired ΔV.

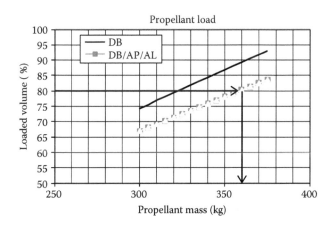

FIGURE 6.23
Propellant load assessment.

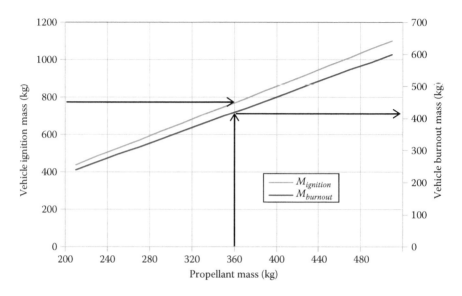

FIGURE 6.24
Vehicle mass analysis.

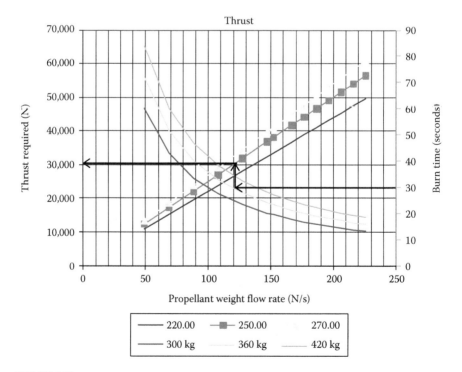

FIGURE 6.25
Burn time and vacuum thrust analysis.

Phase B: Preliminary Design

According to this design analysis, a solid propellant will need to be designed that can deliver 30,000 N of average thrust over 30 seconds to meet the design objective. A simple 1D flyout simulation was developed to test this analysis, and after a few iterations, it was determined that an average thrust closer to 35,500 N would be more likely to meet the ΔV requirement. This was assuming a drag profile computed using the following equation:

$$C_D = C_{D0} + K' \cdot (W)^2 / (Q \cdot S) \tag{6.13}$$

where K' is an empirically determined coefficient, lift is assumed to equal weight (W), and C_{D0} is chosen based on similar configurations of interest with available aerodynamic data. Figure 6.26 presents the results from the first propulsion design iteration.

The thrust–time curve was a table input by hand that attempted to replicate the actual burn profile that might be expected from a neutral burn grain. I_{sp} was set to 250 seconds. Mass flow rate and kinematics (acceleration, velocity, and position) were solved for numerically. The simulation was arbitrarily set to 20,000 m range to go at launch. The performance demonstrates that this design is capable of achieving a Mach 3.5, ΔV, and a flyout range of 20 km in less than 35 seconds. This design could be appropriate for a booster in a long-range interceptor or a boost-glide sustainer in a short-range interceptor.

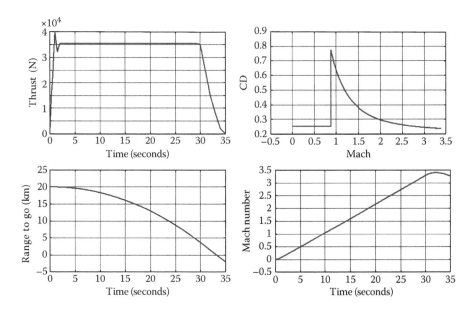

FIGURE 6.26
One-dimensional flyout design results.

6.5.8 Material Properties Design

Material properties design is composed of aeromechanics, aero-heating, and body-bending analysis. Body bending can be assumed under aeromechanics but is a necessary analysis after the design is completed as an input to the flight control design and is best itemized separately. The body bending iteration occurs in Figure 6.16 when materials properties are completed that satisfy the configuration design, the configuration design is then fed back to airframe requirements which in turn is fed back to control requirements.

Mechanical design is usually accomplished with finite element analysis (FEA) techniques. The material properties must meet structural strength requirements set by the maneuverability and boost acceleration design while under the most stressful dynamic pressure and temperature environments. The temperature requirements should be established in the aero-heating analysis. Body temperatures are set by the velocity, which is most likely close to if not exceeding hypersonic speeds, altitude envelope. The trade space is to develop a structure that minimizes weight while surviving the intense mechanical environment. Fleeman [19] provides missile airframe technology assessments that cover hypersonic structural materials including composites and multispectral domes.

In his analysis, Fleeman shows that temperatures are reached above Mach 3 and 10,000 m altitude that eliminates aluminum as a material option. Graphite polyimide, titanium, and steel cover Mach 4–5 while nickel alloys are required above Mach 5. For the AMD mission area, aluminum is not an option at all. Titanium, steel, and nickel alloys will cover the trade space.

Fleeman also covers dome material options that include multispectral (RF–IR) requirements. Dome material scoring is accomplished using dielectric constant, combined midwave and long-wave infrared bandwidths, transverse strength, thermal expansion, erosion resistance, and maximum short-duration temperature. Fleeman selects dome materials that will cover both RF and IR requirements above Mach 3 as spinal/sapphire, quartz-fused silicon, and silicon nitride. Silicon nitride is also selected as a suitable RF only dome for millimeter wave (MMW) seekers at high supersonic and low hypersonic speeds.

6.5.9 Attitude Response Requirements

Attitude response requirements should be satisfied when the guidance and control (G&C) design is combined with the translational response requirements introduced in Figure 6.16. To be clear, when using the term *guidance and control* it should be inclusive of *guidance, estimation, dynamics,* and *control* as these disciplines are not separable in missile systems engineering. Guidance is the cascading of state estimation (sometimes referred to as filtering) and the computation of an acceleration (or some other kinematic state) command using mathematical guidance law(s). Control is the stabilization and behavior shaping of the system dynamics. Developing effective control commands will usually require some form of state estimation. G&C specification and design is performed to satisfy the attitude response requirements of the interceptor design and therefore forms the basis of this section of the book.

Miss distance is the primary interceptor performance requirement to be met in defining and satisfying attitude response requirements. In fact, it is not possible to define attitude response performance requirements without first proposing a kill strategy including defining the warhead lethal radius or requiring hit to kill. Target type(s) and target vulnerability and dynamic characteristics must be assumed before kill strategy can be initially proposed. Seeker tracking error, stability, control saturation, and airframe responsiveness make up the remaining pieces of defining the attitude response requirement. Figure 6.16 shows the proposed translational and attitude response preliminary design approach.

G&C requirements are addressed first. This includes establishing the role physical constraints will play in shaping the preliminary design; the impact dynamic pressure regimes will have on control system selection; and quantifying criteria, namely, miss distance, for computing preliminary design attitude response performance.

The process being proposed will first concentrate on flight control system (FCS) preliminary design. This part of the approach is divided into two pieces, noted as FCS I and II. FCS I is the mechanical piece of the FCS design where steering strategy, actuators, and instrumentation are selected and notably present the hardest challenges as part of the FCS preliminary design in meeting the performance requirements. These designs/selections play the most interactive part with the translational response requirements impacting configuration design, and weight and balance constraints. FCS II is where the controller architecture is defined, designed, analyzed, and simulated. Warren Boord avoids the use of the term *autopilot* as it has many meanings to many engineers. If the reader has interest as to what constitutes the *autopilot*, it would be a combination of the FCS I and II elements into a completed design.

Once the FCS I and II have an initial design and simulation, then it is proposed to move to developing the appropriate set of guidance laws. This process begins with identifying and quantifying the various noise sources and the estimation techniques to mitigate the effects of noise and applying the guidance laws of choice. Note that the choice of guidance laws is plural as this author believes it is a judicious approach to implement more than one guidance law based on a variety of engagement realizations. Guidance loop preliminary design miss distance performance results will need to be evaluated against the kill criteria to select the best set of guidance laws and their implementation architecture given the required engagement realizations. This can be easily done with planar Monte Carlo simulations before integrating the preliminary design into more complete six- or pseudo-six-degree-of-freedom simulation (Figure 6.13). The results of the planar engagement simulation results will either lead to iterations within the FCS design or will lead you back to where requirements flow down began if satisfactory results cannot be achieved.

The attitude response requirements and G&C preliminary design treatment here are not exhaustive documentation as there is an enormous amount of literature on the subject noted by the large number of references associated with this section available to the reader. This section should help the reader focus on the attitude response requirements and a proposed logical process to achieve a reasonable preliminary design. Specific design and analysis approaches will be left to the body of knowledge already available in the literature.

6.5.9.1 Guidance and Control Requirements

The attitude response requirements are met when the combined airframe translational response and the guidance and control (G&C) requirements are met through a consistent preliminary design. Requirement drivers that will flow down from Figure 5.22 are physical constraints, target characteristics, dynamic pressure regimes, and the primary performance metric kill criteria and miss distance. Miss distance, kill strategy, and target vulnerability are used to compute the overall AMD system performance requirement P_{ssk}.

Physical constraints impact G&C design by limiting steering options, as well as limiting the use of, and the size and placement of lifting and stability devices such as wings and fins for control authority. Packaging constraints may indirectly eliminate or limit G&C techniques. Space is needed to house reaction control jets (RCJs) and thrust vector control (TVC) or by having to limit propellant loads shortening control durations to either very early in flight (TVC) or during the last seconds of homing (RCJ). Eliminating control options will impact the preliminary design of the remaining steering options and time constant requirements potentially demanding a trade-off and a revisit of the requirements, the preliminary design, or both.

Target evasive maneuver bounds are the primary contributor to miss distance driving P_{ssk} performance lower and therefore a driving G&C requirement. Target maneuver bounds need to be defined using three performance metrics: maneuver magnitude, period, and command structure [63,76]. These target maneuver metrics must be defined in order to develop effective attitude response and therefore G&C requirements. Evasive maneuver magnitude will stress interceptor kinematic limits; evasive maneuver period will stress interceptor maneuver time constant; and evasive maneuver command structure will stress interceptor agility or acceleration rate-of-change (jerk) performance (t_α) [63,76].

Dynamic pressure regimes will affect responsiveness performance defined here as the rate at which the interceptor airframe will respond to a command signal or the missile time constant (t_α) defined earlier. As engagement altitude increases for a given Mach number time constant performance will degrade and force alternative steering and control strategies to be used to compensate for the loss of control authority.

Phase B: Preliminary Design 141

6.5.9.2 Flight Control Design I

6.5.9.2.1 Steering Policy

Missile steering policy design has several dimensions. The preliminary design steering policy decisions include whether to use skid-to-turn (STT) or bank-to-turn (BTT) maneuver strategy. The STT policy can use a roll attitude or a roll rate control implementation. An STT policy will have equal positive and negative angle-of-attack limitations in both pitch and yaw planes. If a BTT maneuver strategy is used, three options are practically possible. The BTT policy can be a roll to 45°, 90°, or 180° attitude each presenting design advantages, limitations, and implementation challenges. Although there are some potential lift (and thus acceleration) performance advantages to BTT, the literature does not show that for interceptor designs BTT has any clear advantages over STT policy. Moreover, BTT strategies complicate seeker operation and demand increased roll rate bandwidth, and there are multiple concerns over guidance and control system cross-plane coupling that will have to be addressed. In the literature (see, e.g., Xin et al. [44]), there are examples where a combined BTT–STT approach may be preferred.

The implementation of aerodynamic STT tail control is examined first. Plus configurations are rarely if ever used in interceptor class missiles, but for completeness the following is provided. When the missile orientation is at 0° roll, shown in Figure 6.27, we refer to this as the plus configuration. Steering is achieved by a pair of surfaces for pitch and an orthogonal pair of surfaces for yaw.

Strictly considering STT aerodynamic control, an interceptor missile is typically designed with cruciform tail control with lifting surfaces either rotated 45° relative to the controls (interdigitated) or in-line. When the missile is placed in a 45° roll, X-configuration, shown in Figure 6.28, all four panels are deflected for pitch, yaw, and roll steering. This strategy is preferred

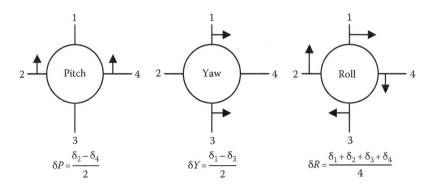

FIGURE 6.27
Definitions of positive pitch, yaw, and roll control for $\phi = 0°$ (arrows show the direction of forces and leading edge of the panel, looking upstream). (From Hemsch, M.J. and Nielsen, J.N., Aerodynamic Considerations for Autopilot Design, *Tactical Missile Aerodynamics*, Vol. 104, Progress in Astronautics and Aeronautics, AIAA, New York, 1986, Chapter 1, Cronvich, L.L. [15].)

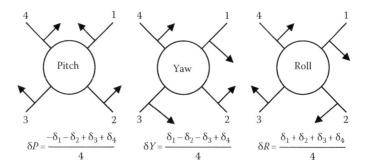

FIGURE 6.28
Definitions of positive pitch, yaw, and roll control for $\phi = 45°$ (arrows show the direction of forces and leading edge of the panel, looking upstream). (From Hemsch, M.J. and Nielsen, J.N., Aerodynamic Considerations for Autopilot Design, *Tactical Missile Aerodynamics*, Vol. 104, Progress in Astronautics and Aeronautics, AIAA, New York, 1986, Chapter 1, Cronvich, L.L. [15].)

over the plus configuration in that more control power for identical surface area can be applied with four fins than with two, for example. Stabilization as well as control is then easier to achieve while demanding that homing commands need to be met in the terminal phase of an engagement.

The aforementioned steering equations are overdetermined in that the commands for three motions (pitch, yaw, and roll) are to be provided by four control surfaces. These equations require another command to make the system determinate. This additional equation, called the "squeeze mode" by Cronvich, Hemsch, and Nielsen [15] (Chapter 1), is chosen so that any axial force resulting from the surface deflections is minimized. A true *squeeze mode* condition shown in Figure 6.29 develops, as missile control surfaces are deflected in such a way that no pitch, yaw, or roll moments exist.

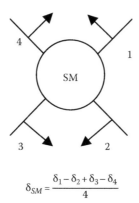

FIGURE 6.29
Squeeze mode tail–fin mixing strategy. (From Hemsch, M.J. and Nielsen, J.N., Aerodynamic Considerations for Autopilot Design, *Tactical Missile Aerodynamics*, Vol. 104, Progress in Astronautics and Aeronautics, AIAA, New York, 1986, Chapter 1, Cronvich, L.L. [15].)

Phase B: Preliminary Design

The tail–fin mixing strategy from Figures 6.28 and 6.29 for combined squeeze mode, pitch, yaw, and roll steering is provided in the following equation:

$$4 * \begin{Bmatrix} \delta P \\ \delta Y \\ \delta R \\ \delta_{SM} \end{Bmatrix} \begin{bmatrix} 1 & 1 & -1 & -1 \\ -1 & 1 & 1 & -1 \\ -1 & -1 & -1 & -1 \\ 1 & -1 & 1 & -1 \end{bmatrix} = \begin{Bmatrix} \delta_1 \\ \delta_2 \\ \delta_3 \\ \delta_4 \end{Bmatrix} \quad (6.14)$$

The squeeze mode command should be chosen to minimize drag, and according to Hemsch and Nielsen [15], chapter 1, control deflection–related drag (D_δ) can be approximated by δ^2. Therefore, commanding zero "δ_{SM}" should be the rule. Hemsch and Nielsen [15], chapter 1, provides an implementation example of tail–fin steering using squeeze mode and is repeated here for convenience. Suppose the commands are $\delta P = 15°$, $\delta Y = 10°$, and $\delta R = 5°$, with $\delta_{SM} = 0$, the resulting deflection angles for each panel are $\delta_1 = 15°$, $\delta_2 = 20°$, $\delta_3 = -5°$, $\delta_4 = -10°$. Computing the approximate control surface-induced drag, $\Sigma\delta_i^2 = 750$ is obtained. Now, suppose the commands are $\delta P = 15°$, $\delta Y = 10°$, and $\delta R = 5°$, with $\delta_{SM} = -5$, the resulting deflection angles for each panel are $\delta_1 = 10°$, $\delta_2 = 25°$, $\delta_3 = -10°$, $\delta_4 = -5°$. Again computing the approximate control surface-induced drag, $\Sigma\delta_i^2 = 850$ results implying increased drag over the previous calculation achieving the same control steering result making zero squeeze mode the preferable trade.

Steering also involves determining whether tail, wing, canard, thrust vector, and/or reaction jet control design strategies will be used singularly or in combination (see, e.g., Nielsen [14] and Wise and Broy [16]). Additional steering strategies, involving thrust vector control (TVC) and reaction jet control (RJC), are employed with aerodynamic tail control to achieve higher angle-of-attack capability and to expand the engagement envelope to higher altitudes and also lower terminal speeds. Multimode steering strategies are one of the required technologies that will enable multimission AMD interceptors.

6.5.9.2.2 Missile Servo Actuators

The basic requirements of an interceptor servo actuator (SA) used for moving the control surfaces in a missile are to operate aerodynamic or thrust vector control steering devices. SA requirements include functionally providing motion of the steering mechanism(s) and with a specified minimum dynamic performance throughout the engagement envelope. The SA must interface with the device being actuated and receive signal commands from the flight controller. The Bode diagram can conveniently measure dynamic performance of the SA. The performance should be stable and specified by gain and phase margins and, moreover, specifying a *no greater than* phase lag in the frequency range of interest. Mechanical hard position limits will be specified (typically no greater than 25°) and a mechanical slew rate limit. Low-frequency noise sources, such as glint, can cause fin-rate saturation leading to flight control system instability. The statistics of these noise sources will have to be specified before SA gain and phase margin requirements can be determined. The aerodynamic hinge moment and the Coulomb friction of the gimbal mechanisms in TVC systems will need to be specified to compute the total load inertia that will lead to a torque specification requirement. Interceptor missile servo actuator requirements can be prioritized as (1) minimum weight/volume and packaging efficiency, (2) shelf life, (3) low cost, (4) reliability, and (5) dynamic performance.

There are many types of SA devices discussed in the literature [45–51] appropriate for interceptor missile applications. The three most common SA types discussed in the literature include pneumatic (cold gas and hot gas types), electrohydraulic (EH), and electromechanical (EM).

Pneumatic devices offer some advantages including torque generation but because they require gas storage or gas generation devices their packaging and weight requirements exceed that can be currently achievable with EM or EH/SA devices. Therefore, they will not be considered further.

Hydraulic actuators are used when large actuation force is required such as with high dynamic pressure regimes and with TVC mechanization. They inherently have long operation time that provides high stiffness and good speed of response that makes them attractive for long-range interceptors and specifically exoatmospheric interceptors. Missiles having *moving wing* configurations would also benefit from hydraulic SA devices, but few interceptor missile designs choose this steering path. Specifically, according to Roskam [46], the EH/SA device offers high natural frequency operation, low electrical power requirements, high power-to-weight ratios, and high power-to-inertia ratios and is available in small to large size and horsepower configurations. Models for EH/SA devices are proposed by Roskam [46], Chapter 10.

The EM SA is most prominently used in modern missile systems. Battery and motor technology has improved dramatically since the inception of missile systems and has enabled their inclusion in designs since the 1970s as discussed by Nachtigal [47] and Goldshine and Lacy [48].

The paper presented by Goldshine and Lacy [48] specifically discusses the SA development for the Standard Missile program. This paper presents a detailed design of a high-performance EM/SA developed by the Pomona Division of the General Dynamics Corporation. This device solved difficult packaging, producibility, and maintainability problems where four-tail surface SA devices were required. The device is specifically a magnetic clutch SA that operates by producing a maximum clockwise (CW) or counterclockwise (CCW) rate-of-change motion to the tail incidence angle. The operating principle depends on maintaining a sustained limit cycle sometimes referred to as a dither or *dithering autopilot*. A detailed diagram, model, and

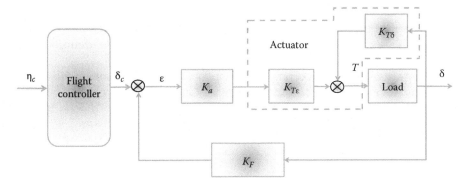

FIGURE 6.30
Generic position servo actuator system functional block diagram. (Adapted from Roskam, J., *Airplane Flight Dynamics and Automatic Flight Controls*, Part II, Roskam Aviation and Engineering Corporation, Ottawa, KS, 1979 [46].)

specifications for this system are provided in the reference. A magnetic clutch SA is also described in detail by Roskam [46].

Figure 6.30 was adapted from Roskam [46] and represents a modeling approach that would be appropriate for either the EH or EM clutch-type SA systems that would be appropriate for preliminary design and requirement development.

More detailed models and analysis techniques are given by Nachtigal [47].

As discussed earlier, there is likely to be a requirement to mix steering strategies to include a combination of RJC, TVC, and aerodynamic control. Wassom and Faupell [45] presents a design and analysis of such an integrated system for high-performance actuation. Their study concluded that using a brushless dc motor, EM/SA device was preferred. The work cited recent technology advances in brushless dc motors, thermal batteries, and power electronics that made this approach feasible and desirable. The system chosen should be able to provide the same power source for the aerodynamic and TVC actuators while solving the long flight time duration requirements that eliminated the pneumatic (both hot gas generator and cold bottle) SA devices. The SA device was shown to perform in a stable manner with a 400° per second slew rate and a 4 ms time constant. This reference shows a detailed schematic, performance specifications, and performance trade-study results. Nachtigal [47] provides detailed models and analysis techniques of brushless dc motor SA systems.

6.5.9.2.3 Instrumentation

Instrumentation is the means to achieving closed-loop automatic flight control. Interceptor accelerations and motions (linear and angular) are measured and/or derived and used for computing error signals and for providing stabilization signals. In this book, an inertial reference unit (IRU) is used for guidance and control and an inertial measurement unit (IMU) is used for navigation. This may not be a universally accepted set of definitions or associations but will work here.

The primary distinction between an IRU and an IMU starts with defining guidance and control and navigation. Guidance and control is a time process of eliminating relative position error between two objects by one of those objects. Navigation is an object precisely establishing its time-dependent location, knowing the precise location of where it wants to end up, and calculating and moving on a course to that final location. Guidance and control requires a real-time measurement of the relative location of its target (seeker) where in navigation the target location is likely to be fixed or slowly moving and predetermined through nonorganic targeting. The relative, time-dependent location precision necessary to complete an intercept is provided by the seeker in guidance and control. It is not provided by an instrumentation package, such as an IMU or an IRU. In navigation, a strapdown or inertially stabilized instrumentation platform provides the high-precision (8–16 significant digits) and highly accurate (low noise and small errors) time-dependent location information. Moreover, navigation instrumentation packages (IMU) are typically updated periodically with satellite navigation systems (global positioning system [GPS] is an example) to retain accuracy over long flights typically encountered by strategic missile systems. This section will not cover the IMU instrumentation but will focus on the IRU used in interceptor guidance and control feedback.

Airframe accelerations and motions are measured by an IRU containing accelerometers, gyros, and resolvers for rolling airframes. The IRU instruments are placed in multiple axis sets, so a complete vector of axial, normal, and lateral kinematic motion, and roll, pitch, and yaw dynamic motion can be computed. This typically, but not necessarily, includes three accelerometers and three gyroscopes arranged in a triad. Table 6.9, GNC Instrumentation Strategy, provides the set of states typically desired for flight control feedback, the instrumentation set used to provide the uncoupled state, and the means used to obtain the state.

TABLE 6.9

GNC Instrumentation Strategy

Feedback State	Instrument Set	Measurement Technique
$\dot{u}, \dot{v}, \dot{w}$	Single accelerometer	Observable
u, v, w	Single accelerometer	Derived
$\dot{p}, \dot{q}, \dot{r}$	Accelerometer set (two or more)	Derived
p, q, r	Rate gyros	Observable
ψ, θ, ϕ	Rate integrating gyro	Observable
α, β	Accelerometer set	Derived

Phase B: Preliminary Design

The IRU location is typically chosen to be close to the axial CG, but due to packaging constraints and the fact that the axial CG travels with the burn of the SRM, it is not usually possible. More important is to place the IRU package in the axis of symmetry.

Three-axis body-fixed accelerometers measure body acceleration ($\dot{u},\dot{v},\dot{w}$) due to the combined aerodynamic, propulsion, and other external forces such as wind gust and ground effects divided by the vehicle mass. Gravitational acceleration or body force is not measured by accelerometers. Assuming the IRU package is not located at the interceptor CG and that the body is symmetric in the Y–Z plane, the accelerometers will sense both kinematic, translational and dynamic, rotational acceleration components. Accelerations sensed by the IRU package, located forward or aft of the CG, are determined from first principles (Newton's second law) and the body reference axis system. Body rotational accelerations, $\dot{\omega}$ ($\dot{p},\dot{q},\dot{r}$), are sensed as a function of the IRU-to-CG offset, r_{IRU}, and coupled body angular rates, ω (p,q,r), are sensed as a result of cross-products (right-hand rule). The generalized equation governing the IRU sensed acceleration is given in the following equation:

$$\eta_{acc} = \eta_{cg} + \dot{\omega} \times r_{IRU} + \omega \times \omega \times r_{IRU} \tag{6.15}$$

A complete coupled, six-degrees-of-freedom set of accelerometer measurement equations that include cg displacement are presented by Abzug [52], Chapter 2, and is repeated here in the following equations for convenience:

$$\eta_{acc\_x} = \frac{F_x}{W} - \frac{\Delta XCG(q^2 + r^2)}{grav} + \frac{\Delta ZCG(\dot{q} + r \cdot p)}{grav} \tag{6.16}$$

$$\eta_{acc\_y} = \frac{F_y}{W} + \frac{\Delta XCG(\dot{r} + p \cdot q)}{grav} + \frac{\Delta ZCG(-\dot{p} + q \cdot r)}{grav} \tag{6.17}$$

$$\eta_{acc\_z} = \frac{F_z}{W} + \frac{\Delta XCG(-\dot{q} + p \cdot r)}{grav} - \frac{\Delta ZCG(p^2 + q^2)}{grav} \tag{6.18}$$

To proceed with preliminary design, it is necessary to have linearized functions to represent the instrumentation and system. Continuous and non-discrete analysis is typically used in preliminary design to develop representative linear transfer functions in terms of Laplace transforms, Nachtigal [47], Chapter 14, to satisfy this requirement.

The literature presumes that a second-order transfer function is representative of an IRU accelerometer arrangement [45–47,54] for all three translational channels. The z-axis accelerometer representation is shown in the following equation:

$$\frac{\eta_{z\_Ha}}{\eta_{z\_M}} = \frac{\omega_a^2}{s^2 + 2 \cdot \zeta_a \cdot \omega_a \cdot s + \omega_a^2} \qquad (6.19)$$

The gyro rate measures the angular rates (p, q, and r) and is also represented by a second-order transfer function as shown in the following equation:

$$\frac{q_{\delta\_Hg}}{q_{\delta\_M}} = \frac{\omega_g^2}{s^2 + 2 \cdot \zeta_g \cdot \omega_g \cdot s + \omega_g^2} \qquad (6.20)$$

Representative values for damping and bandwidths can be found in manufacturer's literature and in the references found in this chapter.

Transfer functions are also necessary to represent linearized equations (6.16 through 6.18) about specific trim conditions (moments are zero at trim not moment slopes) to develop preliminary design results. Later in the design or analysis, state space representations are typically used. Referring back to Table 6.2 and applying the aerodynamic derivative definitions to the flight dynamic equations of motion and following Wise and Broy [54], Figure 6.31 can be developed to represent the linearized z-axis channel. The offset of the IRU package from the c.g. denoted by Δ_{IRU} in Figure 6.31 will shift the zeros of the acceleration transfer function to the right as shown in the following:

$$\eta_{z\_M} = \eta_z + (x_{cg} - x_{IRU}) \cdot \dot{q}_{\delta z} \qquad (6.21)$$

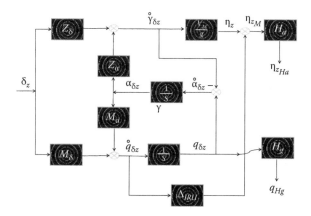

FIGURE 6.31
Uncoupled linearized flight instrumentation measurement.

Phase B: Preliminary Design

The resulting measurement transfer function set input to the IRU is provided in the following equation:

$$\begin{bmatrix} \eta_{z\_M} \\ \delta_z \\ q_{\delta z} \\ \delta_z \end{bmatrix} = \begin{bmatrix} (z_\delta + \Delta_{IRU} \cdot M_\delta) \cdot \dfrac{s^2 + \dfrac{V_M \cdot \Delta_{IRU}}{g}\left[\dfrac{Z_\alpha M_\delta - Z_\delta M_\alpha}{Z_\delta + \Delta_{IRU} M_\delta}\right]s + \dfrac{Z_\alpha M_\delta - Z_\delta M_\alpha}{Z_\delta + \Delta_{IRU} M_\delta}}{s^2 + Z_\alpha \dfrac{V_M}{g} s - M_\alpha} \\ \dfrac{M_\delta s + Z_\delta M_\alpha - M_\delta Z_\alpha}{s^2 + Z_\alpha \dfrac{V_M}{g} s - M_\alpha} \end{bmatrix}$$

(6.22)

The acceleration transfer function from δ_z to η_{zM} has zeros whose magnitudes are a function of IRU placement. When the IRU package is moved forward of the c.g., the zeros go from real to complex and as the IRU is moved to the rear of the c.g. the two zeros are real with one in the right-half plane (RHP). The placement of the IRU can therefore be used to help shape the overall stability and transient performance of the flight controller, and this is why it is important to include this placement in the attitude control preliminary design. Wise and Broy [54] covers how to add TVC and RCS to the design and will not be covered here.

To obtain the flight measurement values out of the instrumentation set, multiply Equation 6.22 by Equations 6.19 and 6.20, respectively. These will be the values used for feedback control in the full-up flight control system.

6.5.9.3 Flight Control Design II

Providing both roll and directional (normal and lateral) control is the highest-level requirement of the flight control system. In this section, the controller architecture piece of the flight control system is covered. The purpose of the controller architecture, including feedback from the IRU, is to provide stability and desirable time history properties while being insensitive to design and disturbance parameter uncertainties (robustness). The controller processes measurements of the current missile states from the IRU sensors with the guidance commands forming a closed-loop FCS and generates new steering commands to achieve the desired states rapidly and in a stable manner. To achieve these top-level flight control system requirements, the process is divided into time and frequency domain requirements that are also typically divided into boost, midcourse, and terminal phases of flight.

Time domain requirements include maintaining closed-loop stability, achieving zero steady-state error, achieving a specified overshoot requirement, and minimizing actuator (position and rate) saturation. Actuator saturation must be avoided throughout the flight phases for any extended period of time, or induced phase lag instabilities will occur in the system and limit the achievable time constants when most needed.

Boost phase will demand/require fast pitch over time constants to accomplish short-range, low-altitude intercepts. High-altitude, longer-range intercepts will be less demanding on boost phase time constant. Exoatmospheric intercepts may require more precise, zero steady-state errors during pitch over control but with longer time constant requirements to achieve flight path angle and velocity to achieve rendezvous (intercept) orbiting requirements. However, requiring minimal overshoot and achieving zero steady-state errors when dynamic pressures are low is more challenging than are endoatmospheric intercepts and typically requires more complex control schemes such as thrust systems.

Midcourse phase will require precise command following or zero steady-state errors through the largest dynamic pressure variations. Low-altitude intercepts may offer the hardest midcourse challenges when nearing earth surfaces by also requiring small overshoots.

Terminal phase will place the hardest demands/requirements on time constant with minimal overshoot and zero steady-state errors. Time constant and overshoot will be competing performance metrics and will require iterative trade-off studies before settling on achievable requirements. Miss distance requirements will ultimately drive these requirements and control configuration requirements.

Frequency domain requirements will take the form of gain and phase margins assuring stability or stability margin and noise or disturbance attenuation properties. Stability margins are required to account for modeling errors and parameter uncertainties. During the flight controller design process, the missile plant is modeled (aerodynamics, mass properties, etc.) with inherent errors and parameter uncertainties from the actual system and especially during varying flight conditions. These deviations from the actual design and during flight conditions present themselves to the controller as phase and gain variations from expected values.

To ensure that system performance requirements are met in light of modeling errors and parameter uncertainties throughout the flight envelope is to design with sufficient stability margins. Modeling errors include neglecting aerodynamic cross-coupling, for example, when designing the controller. Parameter uncertainties include aerodynamic derivative estimate errors, for example. Typical phase margins between 25° and 45° and gain margins between 3 and 12 dB are normally acceptable. It is important to note that designing in margins requires performance trade-offs. For example, increasing gain or phase margin will increase time constants in the expected signal command range. Therefore, it is well worth assessing modeling errors and parameter uncertainties and determining the minimum acceptable margins during preliminary design as requirements.

Phase B: Preliminary Design

To deal with corrupting noise signals and natural disturbances, input frequency attenuation is normally part of the controller design process. Noise sources may include unwanted guidance signals entering the controller, for example. Disturbances will include unmodeled bending modes of the airframe and wind gusts, for example. These noise and disturbance sources will present themselves as having high frequency when compared to wanted command signals and are fairly easy to deal with. Frequency attenuation is normally specified as part of the controller requirements where there will be a minimum rate of gain roll-off per decade of frequency. Moreover, there should be a requirement specifying the amount of attenuation required at a specific frequency. The trade space is command sensitivity with frequency attenuation. This will create some difficulty in the terminal phase where engagement states change quickly and these legitimate higher-frequency demands will require interceptor response.

The modern controller is a digital computer programmed with a system of equations that translates guidance command signals (acceleration, rate, or position demands in the three dynamic planes of missile motion) to steering actions (steering system deflections or actuations) to achieve the desired state set points meeting the requirements discussed earlier.

The preliminary design process begins with a complete characterization of the airframe aerodynamics either in a single plane for axisymmetric missiles or in two separate directional planes for nonaxisymmetric missiles assuming the roll plane is ignored. The full-up six-degree-of-freedom aerodynamic model is linearized (see Bar-On and Adams [57]) and trimmed for specific flight conditions. Usually, the edges of the flight envelope and select points in the heart of the flight envelope for each flight phase are selected for trimming. Trim points at selected Mach, angle-of-attack, and dynamic pressure combinations are resolved into either classic transfer functions or modern state space models or both (recommended). The missile states typically chosen to be controlled or observed include differential or perturbed (from trim state) forward velocity (u'), angle of attack (α'), pitch rate (q'), flight path angle (γ'), and normal/lateral acceleration (η_z). Acceleration is the missile state that is most likely being commanded. This architecture is then referred to as an acceleration command flight control system. If Figure 6.31 is rearranged and it is assumed that the IRU measurements are unity, then one possible three-loop FCS configuration or architecture for preliminary design is provided in Figure 6.32. In this architecture, a simple gain, K, is used as the controller. It can be replaced by some other controller configuration after a satisfactory gain is found. For example, the gain, $K(s) = k_p(s + (k_i/k_p))/s$, is selected for the case of a proportional-plus-integral (PI) controller.

Guidance generates an acceleration command and the primary feedback loop provides achieved airframe acceleration. They are combined as shown in Figure 6.32, forming the acceleration error signal, e_1. The angle of attack and pitch rate are combined to form the inner loop and are used for stability purposes.

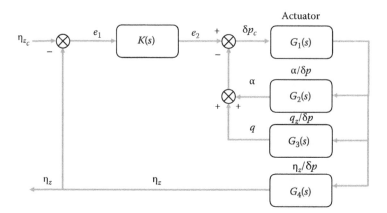

FIGURE 6.32
A simple missile flight control system architecture.

Transfer functions representing each state of interest in the controller solution are shown and labeled accordingly:

$$G_1(s) = \frac{K_{act}}{\tau_{act}(s)+1}; \quad G_2(s) = \frac{K_3[-K_4 \cdot s + 1]}{\omega_{af}^2 \cdot s^2 + \frac{2\zeta_{af}}{\omega_{af}} \cdot s + 1}; \quad G_3(s) = \frac{K_1[T_\alpha \cdot s + 1]}{\omega_{af}^2 \cdot s^2 + \frac{2\zeta_{af}}{\omega_{af}} \cdot s + 1};$$

$$G_4(s) = \frac{\frac{-K_1}{k_{grav}}[-K_2 \cdot s^2 + 1]}{\omega_{af}^2 \cdot s^2 + \frac{2\zeta_{af}}{\omega_{af}} \cdot s + 1}$$

Forming the following equivalents,

$$A = \frac{\tau_{act}}{\omega_{af}^2}; \quad B = \frac{1}{\omega_{af}^2} + \frac{2 \cdot \zeta \cdot \tau_{act}}{\omega_{af}}; \quad C = \tau_{act} + K_{act}(K_1 \cdot T_\alpha - K_3 \cdot K_4) + \frac{2 \cdot \zeta}{\omega_{af}};$$

$$D = 1 - K_{act}(K_1 + K_3); \quad E = (-K_2)$$

and assigning

$$A' = \frac{A}{D}; \quad B' = \frac{B}{D}; \quad C' = \frac{C}{D}$$

the remaining coefficients and terms are defined as

$$K_1 = \frac{Z_\delta \cdot M_\alpha - Z_\alpha \cdot M_\delta}{M_\alpha}; \quad K_3 = \frac{M_\delta}{M_\alpha}; \quad K_4 = \frac{-Z_\delta}{M_\delta}; \quad T_\alpha = \frac{M_\delta}{Z_\delta \cdot M_\alpha - Z_\alpha \cdot M_\delta}$$

Phase B: Preliminary Design

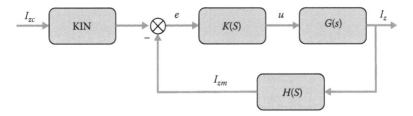

FIGURE 6.33
Simplified missile flight control system architecture.

Then, the open-loop plant transfer function with α and q feedback can be defined as $G(s)$ in the reconstructed control architecture in Figure 6.33, where

$$G(s) = \frac{\dfrac{K_{act} \cdot K_1}{D \cdot kgrav}\left[-K_2 \cdot s^2 + 1\right]}{A' \cdot s^3 + B' \cdot s^2 + C' \cdot s + 1} \tag{6.23}$$

For the first iteration assume $H(S)$ is unity. $K(S)$ will be determined by the root locus method to adjust the time history properties, and KIN will be used to set the steady-state error to zero.

The root locus shown in Figure 6.34 indicates that a stable closed-loop system can be achieved when $0.01 < K < 2.7$. The lower limit, K, produces a sluggish design, while the upper limit is too oscillatory with long settling times and large overshoot.

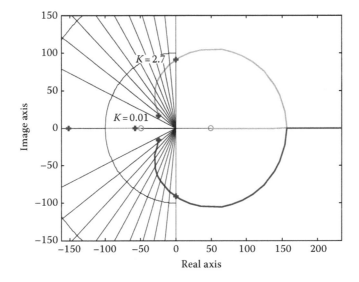

FIGURE 6.34
Missile flight control system root locus gain sensitivity.

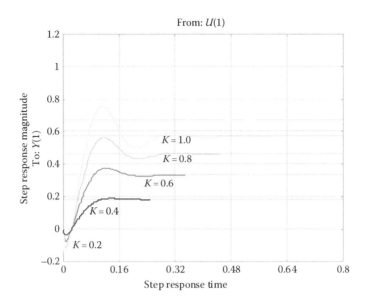

FIGURE 6.35
Missile flight control system closed-loop step response and gain sensitivity.

The first iteration attitude requirement set is selected to achieve a desired zero steady-state error with a time constant $\tau < 0.1$ seconds, overshoot of <2%, and a settling time <0.2 seconds. A gain margin $5 < GM < 10$ and a phase margin $PM > 30°$ are desirable. Figure 6.35 overlays the step response time histories for $0.2 < K < 1.0$.

Selecting $K = 0.4$ satisfies all of the first iteration preliminary design required performance parameters as shown in the associated Bode plot and step response time history in Figures 6.36 and 6.37, respectively.

The next step in this design is to combine the FCS preliminary design with the guidance law(s) of choice in a nonlinear simulation and establish whether these requirements will allow the other performance requirements to be satisfied.

6.5.9.4 Guidance Law Design

The missile guidance law preliminary design and performance evaluation begins with a flow-down of kill criteria requirements and results from the preliminary seeker design and performance. Once these requirements and inputs are established, combined linear and nonlinear techniques and Monte Carlo modeling and simulation are employed. Adjoint analysis approaches [58] have historically played a significant role in this phase of design since the 1950s, but modern computational capability has enabled a more accurate Monte Carlo modeling and simulation on inexpensive desktop and laptop

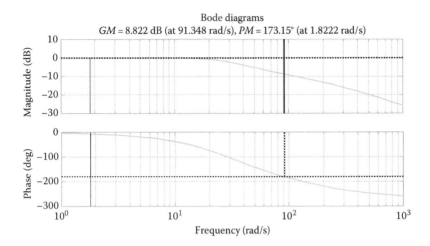

FIGURE 6.36
Missile flight control system Bode plot and margins for $K = 0.4$.

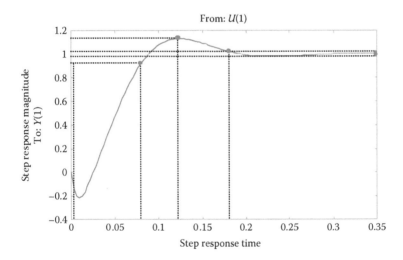

FIGURE 6.37
Missile flight control system closed-loop step response ($K = 0.4$).

computers and is most recommended. In the Monte Carlo approach, guidance loop elements are often represented with transfer functions and simplified mathematical expressions. Figure 6.38 is a nonlinear guidance loop simulation block diagram with both linear and nonlinear representations of various functional components used for design and trade-off studies necessary to meet the principal objective of hitting and/or destroying the required target sets.

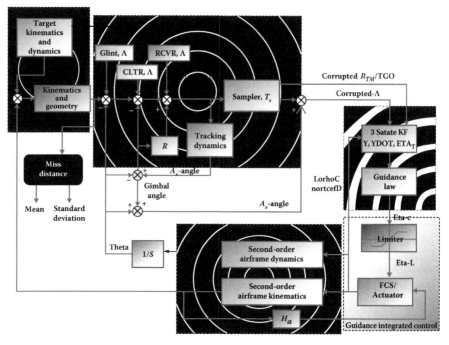

FIGURE 6.38
Homing loop preliminary design model block diagram.

The missile guidance preliminary design problem, shown in yellow, is to develop a guidance law or a set of guidance laws and filtering approaches that successfully complete the terminal homing engagement. From a requirements standpoint, this means satisfying the miss distance (or hit-to-kill) requirement that flows down from the single-shot probability-of-kill, P_{ssk}, top-level requirement with the imposed constraint of kill strategy and seeker preliminary design performance against a specific target set or sets. In following iterations, these constraints may need to be relaxed, modified, or changed after the first, and possibly subsequent, terminal homing guidance preliminary design phase(s) is completed.

There are four primary guidance law preliminary design constraints, shown in blue, to consider while addressing the primary, miss distance, design driver requirements. Specifically, guidance law design is achieved through properly characterizing, modeling, and simulating the function and performance of the guidance integrated target (GIT) factors, guidance integrated noise (GIN) factors, guidance integrated control (GIC) factors, and guidance integrated airframe (GIA) factors.

Modern guidance design is primarily based on modern estimation and control theory explicitly dealing with time-varying systems and nonstationary noise processes. The literature is exhaustive providing various modern guidance law design approaches and law performance analysis approaches (e.g., Stewart and Smith [58], Stallard [59], Yanushevsky [63], Crassidis and Junkins [64], Gurfil et al. [65], Guelman [66,67], Adler [68], Cochran et al. [69], Song [70], Yang and Yang [71], Yuan and Chern [72], Yueh [73], and Biggers [74]), and they are not repeated here. Most modern guidance design and analysis approaches employ the separation theorem (e.g., Stallard [59]) optimizing a linear guidance system quadratic performance index (e.g., mean-square miss distance) with a constraint equation where an integral-square interceptor lateral acceleration is specified. Employing the separation theorem permits the G&C problem to be divided into two subordinate problems of optimal estimation and optimal control.

Optimal estimation here is synonymous with Kalman (Kalman–Bucy) filtering [59] and is employed to produce measured state estimates of the target signal in the presence of noise, which are then acted upon by the guidance law to produce acceleration commands. Specifically, optimal estimation applies Kalman filtering to corrupted LOS, range, and their rate measurements. The GIN is assumed *white*, and the filtering solution is obtained from calculating the ensemble minimum mean square error of noise plus signal.

Optimal control is employed to minimize a performance index (e.g., mean-square miss distance) to produce time-dependent guidance law gains used to compute terminal homing commands necessary to close an angle and range on a dynamic target. A classical guidance law design produces fixed gains. Specifically, optimal control addresses the GIT, GIC, and GIA constraints including maintaining stability and minimizing system saturation where angular rate, position, or force limits are reached before achieving the desired guidance commands. As the time to go decreases, control saturation design issues become hard to solve and are best handled with time-varying gains. The GIA constraint, airframe time constant, is a primary design trade that will be in the trade space. The airframe time constant trade may have to be satisfied by adjustments to configuration, mass and inertia properties, and control effectiveness.

The next section covers detailed GIN sources, which are necessary to characterize, model, and simulate a complete guidance law preliminary design.

6.5.9.4.1 Guidance Integrated Noise

The guidance integrated noise (GIN) blocks shown in Figure 6.38 represent important design and performance constraints that corrupt the interceptor-to-target line-of-sight (LOS, λ) measurements. The LOS rate measurement, typically measured within seeker tracking, is accomplished in two perpendicular axes and relative to the interceptor antenna (A_x) or optics that tracks the target motion. The LOS measurement is corrupted by a variety of noise sources including glint, receiver (RCVR), and clutter (CLTR) noises that are characterized by their power spectral densities (PSD, Φ), standard deviation (STD, σ), and their correlation time constants (T_N). In general, these noise sources are categorized as range dependent or independent and can be

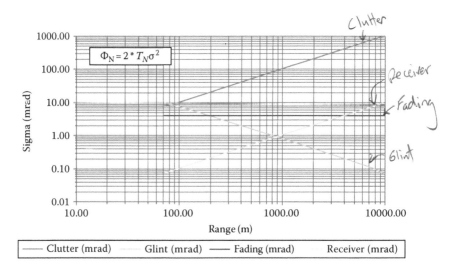

FIGURE 6.39
Angular noise sources and representative standard deviations.

represented as shown in Figure 6.39. Fading noise is range independent and is usually easiest to filter and is not shown in Figure 6.38.

Other sources of GIN not shown are multipath error, radome boresight error (RBE), and handover error. An attempt should be made to add these error sources into the guidance law design process individually and in combination.

Multipath presents angular error due to signal reflections from surfaces being mistaken for the valid target signal. Multipath is therefore most important for look-down engagements described in the literature [60–62] and can be difficult to compensate or filter.

Radome boresight error (RBE, "R") is defined as the rate of change of the refraction angle of the target signal with a change in the gimbal angle and is a function of radome characteristics including material type, physical dimensions, and material signal bandwidth. RBE acts to destabilize the airframe that adversely affects the missile aerodynamic design including wing size and body characteristics. Representative RBE may vary between 0.01 and 0.025 deg/deg.

Handover error comprises the heading error (HE) and time-to-go estimation error. Handover error is another source of GIN that is not shown explicitly in Figure 6.38 as it is typically accounted for in midcourse guidance; however, it could just as well be another added noise component to the LOS error. HE is the angular error measurement between the solution to the target-to-missile collision triangle and the missile velocity vector direction when terminal homing begins. Time-to-go estimation errors are the result of error on the range-to-go estimate and a combination of target and own missile velocity estimation errors. The mathematics associated with handover error was given in Chapter 5.

The *Sampler*, T_s block in Figure 6.38 is used to account for the receiver data collection and processing time and is a combination of a sample hold and processing delay. These values may vary between 10 and 50 ms for the sample hold and an additional 10–25 ms for a processing delay. Significant and comprehensive detail can be found on all of the GIN sources and how they are represented in modeling and simulation in the referenced literature.

6.5.9.4.1.1 Guidance Integrated Target GIT refers primarily to the attacking target dynamics and is the primary miss distance driver of the engaging interceptor. Evasive maneuvers are one, if not the most, effective defense penetration feature (DPF) used on offensive missiles [59,61,63,72–74]. The evasive maneuver exploits the terminal homing portion of the interceptor and is typically the weakest link of the defense system attacking the engagement weapon aerodynamic time constant, potentially causing intolerable miss distances for hit-to-kill missiles/guns or causing increased miss distances that are unacceptable for warhead kill missiles. Therefore, it is of primary importance to properly characterize, model, and simulate the GIT for a guidance law preliminary design.

Target maneuver constitutes any kinematics and/or dynamics state changes in magnitude and/or direction. Regardless of the motivation of the maneuver, from a design standpoint, the target maneuver acts to evade an intercepting missile by inducing miss distance. If designed properly, an evasive maneuver can render the entire defense system ineffective despite whether or not all of the other defense system elements are doing their job. A properly designed evasive defense penetration maneuver is *the maneuver that causes the defensive weapon to miss with sufficient distance that no damage is caused to the evading target while yet not prohibiting target from accomplishing its mission.*

Evasive maneuver design parameters include weave magnitude, weave period (for a weaving evasion) initiation time, and duration. The preliminary flow-down design requirement is to minimize the effect of practical evasive maneuvers, and therefore, the bandwidths of potential evasive maneuver design parameters are important inputs to the guidance and control preliminary design.

Unfortunately, it is difficult to build a general analytical model which is well matched to reality and that can be utilized during design. Statistical models are most readily used [58,59,63,64,72,74] and have their place in the design process but in general lack the dynamic or physics-based limitations imposed on the evading missile that are typically imposed on the intercepting missile in the form of time constants. The design results are then pessimistic from the weapon designer's perspective leading to potentially unnecessary trades and compromises. To design the estimator, the target is modeled using a statistical process approach, but when implemented within the homing model, the target will need to be a more accurate representation.

A more appropriate preliminary design approach concerning target evasive maneuver is to characterize the requirements using a two-part process. The first part is to establish the maneuver design permissive bounds (MDPBs) of potential offensive target designs or the maneuver magnitude limit requirement in a capability-based acquisition. The second part is an optimization problem. Formally, it should first be determined what target magnitude capability should the defensive system be capable of successfully engaging and then what would be the optimal maneuver characteristics given a target class.

The MDPB objective is to bound the maximum achievable, range-sensitive maneuver magnitude–period combinations. These combinations are a function of initiation and duration times and are variable with dynamic pressure. Next, interceptor miss distance is significantly influenced by evasive maneuver weave frequency, and it can be shown that there exists a frequency for which the amplitude of the miss is a maximum [63].

The preliminary guidance law design and performance analysis should proceed using identically commensurate dynamics for both the evader and pursuer. A reasonable first approach is to employ a low-order equivalent model (LOEM) in the homing loop model characterized in second order by the airframe natural frequency and damping and a first-order flight control system time constant. Tail-controlled missiles should include the right-half plane zero (nonminimum phase systems). An example LOEM is given in the following equation:

$$W(s) = \frac{1 - \dfrac{s^2}{\omega_z^2}}{(1 + \tau s)\left(1 + \dfrac{2\zeta}{\omega_{af}} s + \dfrac{s^2}{\omega_{af}^2}\right)} \qquad (6.24)$$

Accordingly, a closed-form solution for miss distance can be obtained by Yanushevsky [63] as a function of the guidance law under study and the effective navigation parameters, guidance system time constant, natural frequency, and damping ratio to determine the optimal weave frequency for which the amplitude of the miss distance has a maximum. The established existence and the procedure of determining the miss distance maximizing weaving frequency offer the optimization approach to the design of the worst-case scenario when developing defensive missiles to defeat maneuvering targets. Figure 6.40 shows an example of the results for determining the optimal evasive maneuver weave frequency for an interceptor modeled with the given parameters [63].

The results from this phase are inputs for preliminary design utilizing a planar 2D Monte Carlo model. The results from the planar model and

Phase B: Preliminary Design

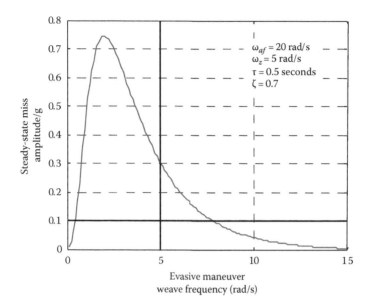

FIGURE 6.40
Example illustrative optimal evasive maneuver weave frequency results on miss distance.

simulation once an apparent acceptable preliminary design is achieved will be made more precise by utilizing detailed nonlinear six-degree-of-freedom engagement models.

6.5.9.4.1.2 Guidance Integrated Control The GIC module receives the guidance law acceleration command and represents the flight control system (FCS) response. It is important to represent the FCS time delay and time-dependent response magnitude accurately. This is best accomplished using transfer functions. The nonlinear limiting of the command and response must also be included in this modeling to pick up saturation levels and realistic airframe limits.

6.5.9.4.1.3 Guidance Integrated Airframe The GIA is represented by a pair of transfer functions for the missile body angular rate response and the missile c.g. acceleration response to control deflections. It is in the *second-order airframe dynamics* transfer function that an accurate representation of airframe time constant is required. The *second-order airframe kinematics* block must represent both the normal/lateral interceptor c.g. acceleration and its axial acceleration or slow-down effects. Oftentimes, airframe slow-down is neglected in guidance law design and analysis, which is a mistake. Slow-down is a major contributor to miss distance and is therefore an important parameter in designing the guidance law and its implementation. Airframe slow-down within the pitch plane can be represented in a simplified fashion as shown in the following equations:

$$\dot{V}_M = \left[\frac{Thrust - D_{ZL}}{W}\cos\alpha - \eta_M \sin\alpha\right] \quad (6.25)$$

$$\dot{\gamma} = \frac{9.81}{V_M}\left[\frac{Thrust - D_{ZL}}{W}\sin\alpha + \eta_M \cos\alpha\right] \quad (6.26)$$

The term D_{ZL} represents the zero-lift drag and not axial force and η_M represents normal (or lateral when using β) acceleration in g's. The angle of attack, α, can be represented as

$$\alpha = \tau_\alpha \cdot \dot{\gamma} \quad (6.27)$$

The airframe time constant (τ_α) equation was given in Table 6.2.

Homing analysis when high-frequency guidance signals are expected requires a model of the interceptor body-bending modes. Hemsch and Nielsen [15] provides a detailed treatment of how to handle missile body-bending modes in flight control system analysis and could apply for preliminary design. Guidance law preliminary design does not necessarily require this analysis unless during a particular design trial it is discovered that high-frequency commands are necessary. If this is the case, see Hemsch and Nielsen [15].

6.5.7.9.1.4 Estimation As discussed in the preceding sections, the noise sources, disturbances, and modeling inaccuracies corrupt the measurements that are intended for use in the guidance law design and performance analysis. To prepare the *preliminary design problem* of Figure 6.38, an estimator needs to be designed. This estimator will no doubt be iteratively changed in the preliminary design process. The Kalman–Bucy filtering (KBF) approach is a commonly used filter and will be implemented here. KBF is based on a probabilistic treatment of process and measurement noise sources, and therefore, the probability theory is used to model the corruption and uncertainty sources in a Monte Carlo sense described earlier. Essentially, the KBF is used to reconstruct the measured states from the noisy measurements. The KBF is usually a good choice for guidance applications because it is fundamentally a low-pass filter where legitimate guidance measurements will have a relatively low frequency while unwanted disturbances and noise typically have a high frequency. The literature on this subject is exhaustive and should be consulted for implementation.

6.5.9.4.1.5 Mathematical Guidance Law Design The context of this discussion on missile guidance law design is within the terminal homing portion of the engagement. It is assumed that a midcourse guidance law was implemented and handover errors exist during the transition from midcourse to terminal homing and that some relatively short terminal homing time period remains to zero-out miss distance. Based on terminal engagement encounter conditions (requirements) and the flow-down of requirements, a mathematical guidance law will have to be selected to satisfy the kill criteria. The guidance law is selected based on a series of trade-offs between AMD system engagement envelope and P_{ssk} performance requirements and the cost and complexity of implementation. The objective of the guidance law is to mathematically produce commands from relative target-to-interceptor state measurements that are translated into actuator control signals that steer the interceptor into a collision course with the target while minimizing the amount of energy expended. It is likely that more than one guidance law algorithm will be required to fulfill the interceptor engagement encounter envelope requirements.

The most widely used terminal guidance law is proportional navigation (PN). PN is the optimal intercept solution minimizing the minimum mean-square miss distance plus the weighted integral square interceptor acceleration perpendicular to the LOS vector for a constant closing velocity, nonmaneuvering target. PN was established as a guidance law in the 1940s and received the focus of attention when intercept missiles became prevalent in the 1950s, and since this time, there has been no other single guidance law that is the center of continued practical research and development. Today, PN is the established guidance law benchmark for both nonmaneuvering and maneuvering target engagements.

The applied principle of PN is that a mathematically generated acceleration command steers the interceptor velocity vector to nullify the interceptor-to-target LOS rate. PN performance varies with choices of design parameters including steering gain, navigation gain, and additive terms, all of which may or may not change with time to go. Moreover, either linear or nonlinear implementations of PN may be used singularly or in combination. When PN is solely dependent on relative angle rate information, it is easy to mechanize but is directly tied to the seeker angle tracking mechanization and the associated tracking errors. As such, the PN algorithm design will be closely tied to the angle tracking error and error mitigation design approaches. The selection of these design parameters will be discussed in the remainder of this section.

Much of the PN design work in the last 40 years has been to improve performance against maneuvering targets and in variable closing velocity engagements. PN implementations fall in one of two classes [65–67], either the true proportional navigation (TPN) or the pure proportional navigation (PPN) law class. TPN generates a maneuver acceleration command perpendicular to the instantaneous LOS. PPN generates a maneuver acceleration perpendicular to the instantaneous velocity vector. The difference between TPN and PPN is in the calculation of the effective navigation constant N' in the following equation's PN command representation:

$$\eta_c = N' \cdot V_c \cdot \dot{\lambda} \tag{6.28}$$

where

$$N' = \begin{cases} N \cdot V_M \cdot \cos(\gamma_M - \lambda) / V_C \Rightarrow PPN \\ N \cdot V_M / V_C \Rightarrow TPN \end{cases} \tag{6.29}$$

The effective navigation constant can take on values $2 \leq N' < 5$ to generate a collision course typically without violating stability requirements.

As part of the preliminary design process, maintaining stability throughout the engagement is an added criterion. As the range to go decreases, the problem of maintaining guidance law stability severely increases and will adversely affect miss distance and therefore is a primary design driver. Within the context of preliminary design when dealing with a roll attitude–stabilized missile, it is reasonable to assume that the directional planes are decoupled and can be modeled independently to accomplish the objectives, and typically, this design work is accomplished in decoupled planes. Therefore, due to its simplicity, much of the work on PN has been accomplished in single plane analysis. Arguably, using a 3D PN preliminary design approach will produce different and more accurate results and may be appropriate for the later stages of preliminary design. The mathematical model for the 3D engagement with PN is cumbersome but can be found in the literature (see, e.g., Adler [68], Cochran et al. [69], Song [70], and Yang and Yang [71]).

Other linear and nonlinear forms of PN have been produced in the literature to address some of the inherent vulnerabilities in the PN guidance law assumptions to include constant closing velocities, nonmaneuvering targets, and short-range instability. These vulnerabilities in the law have been the focus of much research and have produced improvements to the original classical PN law. For example, an additional PN approach, realistic true proportional navigation (R-TPN), has been proposed in the literature [72]. R-TPN requires the use of active radar Doppler processing to measure the effects due to nonconstant closing velocities. These closing velocity measurements are incorporated into the law through time-varying gain adjustments.

When considering improvements regarding maneuvering targets, there have been many new PN formulations. One of the linear forms of PN developed to address the maneuvering target is augmented proportional navigation (APN). APN is the optimal control law, where the miss distance criteria go to zero having a minimum integral-squared interceptor acceleration constraint, for a noise-free, zero-lag guidance system against a target executing

Phase B: Preliminary Design

an instantaneous step maneuver perpendicular to the interceptor-to-target LOS. The APN is written in Equation 6.21:

$$\eta_c = N' \cdot V_c \cdot \dot{\lambda} + \frac{N' \eta_T}{2} \quad (6.30)$$

The step target maneuver in the previous equation is represented by η_T.

The APN has shown promise to improve homing performance in some limited cases against maneuvering targets as would the theoretical development of the law suggest. APN homing performance improvement over PN is tightly tied to target maneuver estimation performance and the minimization of guidance lags. One such APN practical implementation is presented by Yueh [73].

The most practical and useful way to address maneuvering targets and to account for guidance system lags is to modify the PN guidance law by adding lead and nonlinear PN terms. The lead term in the modified law will inherently reduce the effects of the dominant guidance system lag terms, while the nonlinear term(s) if selected properly can be used to weigh the error signal, most notably LOS rate in PN, to reduce the effects of highly evasive target maneuvers that are typically oscillatory in nature. Neither guidance system lags nor oscillatory maneuvers are assumed in the APN law derivation.

The LOS rate measurement from the angle track loop can be modified with a lead term filter as shown in Figure 6.41.

The resulting transfer function is provided in Equation 6.31, where τ_A ($\tau_A = 1/K_1$) is the angle track loop time constant:

$$\frac{v_s}{\dot{\lambda}} = \frac{\tau_2 s + 1}{\tau_A \cdot \tau_1 \cdot \tau_2 s^3 + \tau_A \cdot (\tau_1 + \tau_2) s^2 + \tau_A \cdot s + 1} \quad (6.31)$$

The trade space with this filter has higher sensitivity to radome boresight error and premature loss of stability and is more sensitive to glint or scintillation, increasing miss distance due to these sources. The parameters τ_1 and τ_A are chosen based on S:N and τ_2 will be small so as to improve damping

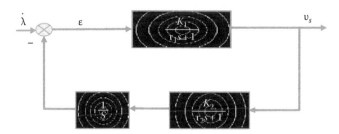

FIGURE 6.41
Lead term addition to LOS rate measurement.

due to noise. K_2 is 1 or greater. The value υ is used as the noisy LOS rate and passed to the Kalman filter for estimation.

Nonlinear PN was investigated by Biggers [74] as early as 1969, to improve the capabilities of PN to countertarget evasive oscillatory maneuvers. Biggers offered the following equation as PN with a nonlinear modification term without any mathematical derivation:

$$\dot{\gamma} = N'_1 \dot{\lambda} + N'_n \dot{\lambda}^n \tag{6.32}$$

When the previous equation is multiplied by V_C, it returns the usual acceleration command of pure or true PN. The new guidance law proposed that "n" be an odd integer. By inspection when the LOS rate is large, it is weighted more heavily, and as it approached zero, the desired condition, it nearly disappears and defaults to the traditional PN. The nonlinear term causes large LOS rates to be corrected rapidly. Preliminary results obtained by Biggers show marked miss distance improvement against sinusoidal maneuvering targets employing Equation 6.32 when $n = 3$ over traditional PN.

A rigorous derivation and new nonlinear guidance (NLG) law design approach was offered by Yanushevsky and Boord [75] where the optimality of the cubic LOS rate term was proven using Lyapunov control techniques and additionally proves the optimality of the well-known PN guidance law against nonmaneuvering targets. The new NLG laws developed by Yanushevsky and Boord [75] improve interceptor homing performance against maneuvering targets when compared to PN and APN guidance laws without requiring any additional information from the sensor. The nonlinear guidance laws take the form shown in Equation 6.33. The first term is the classic PN term, the second term (third solution) is nonlinear, and so on:

$$\eta_c = N \cdot v_c \cdot \dot{\lambda} + \sum_{k=1}^{3} u_k \Rightarrow k = 1, 3, 5, 7, \ldots$$

$$u_1 = N_1 \cdot \dot{\lambda}^3(t) \tag{6.33}$$

$$u_3 = N_3 \cdot a_T(t)$$

A 3D derivation of this NLG law is provided by Yanushevsky [76].

References

1. Warren, R.S., Price, C.F., Gelb, A., and Vander Velde, W.E., Direct statistical evaluation of nonlinear guidance systems, in *AIAA Guidance and Control Conference*, Paper No. 73-836, Seattle, WA, August 20–22, 1973.

2. Calamia, M., Jiberio, R., Franleschetti, G., and Givli, D., Radar tracking of low altitude targets, *IEEE Transactions on Aerospace and Electronic Systems*, AES-10, 539–544, 1974.
3. Nesline, W.F., Missile guidance for low altitude air defense, *AIAA Journal Guidance, Control, and Dynamics*, 2(4), 283–289, July–August 1979.
4. Tsuchiya, T. and Takashige, M., Optimal conceptual design of two-stage reusable rocket vehicles including trajectory optimization, *Journal of Spacecraft and Rockets*, 41(5), 770–7789, September–October 2004.
5. Fleeman, E.L., *Tactical Missile Design*, AIAA Education Series, 2nd edn., AIAA, Reston, VA, July 2002.
6. Fleeman, E.L. and Donatelli, G.A., *Conceptual Design Procedure Applied to a Typical Air Launched Missile*, AIAA-81-1688, Dayton, OH, August 11–13, 1981.
7. Chin, S.S., *Missile Configuration Design*, McGraw-Hill, New York, 1961.
8. Blake, W.B., Missile DATCOM, AFRL-VA-WP-TR-1998-3009, February 1998.
9. Moore, F.G. and Moore, L.Y., The AP09, Report Number A365574, January 2008.
10. Nesline Jr., W.F. and Nesline, M.L., Wing size versus radome compensation in aerodynamically controlled radar homing missiles, AIAA Paper 85-1869, Reston, VA, 1985.
11. Moore, F.G., *Approximate Methods for Weapon Aerodynamics*, Progress in Astronautics and Aeronautics Series, V-186, AIAA, Reston, VA, 2000.
12. Slack, A.J., Unpublished work and presentation, Battlespace engineering assessment tool, Technology Service Corporation, Silver Spring, MD, 2002.
13. Krieger, R.J. and Williams, J.E., High performance missiles of the future, McDonnell Douglas Astronautics Company, Neilson Engineering Research, St, Santa Clara, CA, 1988.
14. Nielsen, J.N., *The Present Status and the Future of Missile Aerodynamics*, NASA Technical Memorandum, Moffett Field, CA, January 1988.
15. Hemsch, M.J. and Nielsen, J.N., Aerodynamic Considerations for Autopilot Design, *Tactical Missile Aerodynamics*, Vol. 104, Progress in Astronautics and Aeronautics, AIAA, New York, 1986 (Chapter 1, Cronvich, L.L.).
16. Wise, K.A. and Broy, D.J., Agile missile dynamics and control, *AIAA Journal of Guidance, Control and Dynamics*, 21(3), 441–449, May–June 1998.
17. Sutton, G.P. and Biblarz, O., *Rocket Propulsion Elements*, 7th edn., John Wiley & Sons, New York, 2001.
18. Briggs, M.M. and Tahk, M.-J., Appropriate aeroprediction accuracy in defining missile systems for aerodynamic flight, in *NEAR Conference on Aerodynamics*, Paper No. 4, Neilsen Engineering and Research, Monterey, CA, 1988.
19. Fleeman, E.L., Technologies for future precision strike missile systems-missile aeromechanics technology, in presented at *the RTO SCI Lecture Series on Technologies for Future Precision Strike Missile Systems*, RTO-EN-018, Paris, France, June 18–19, 2001.
20. James, D.A., *Radar Homing Guidance for Tactical Missiles*, Macmillan Education, Basingstoke, UK, 1986.
21. Maksimov, M.V. and Gorgonov, G.I., *Electronic Homing Systems*, Artech House, Norwood, MA, 1988.
22. Edde, B., *RADAR Principles, Technology and Applications*, Prentice Hall, Englewood Cliffs, NJ, 1993.
23. Barton, D.K., *Radar System Analysis*, Prentice Hall, Englewood Cliffs, NJ, 1964.

24. Weidler, R.D., Target LOS and seeker head equations for an AAM simulation, Flight Dynamics Group, Stability and Flight Control Section, FZE-977, November 19, 1969.
25. Hendeby, G., Development and evaluation of an active radio frequency seeker model for missile with data-link capability, LiTH-ISY-EX-3309-2002, December 11, 2002.
26. The George Washington University, School of Engineering, Modern air-to-air radar systems, A short course, Washington, DC, April 26–29, 1983.
27. Nathanson, F.E. and Jones, W., Airborne radar, an intensive short course, Technology Service Corporation, Silver Spring, MD, Circa 1980.
28. Schleher, D.C., *Introduction to Electronic Warfare*, Artech House, Dedham, MA, 1986.
29. Miwa, S., Imado, F., and Kuroda, T., Clutter effect on the miss distance of a radar homing missile, *AIAA Journal of Guidance, Control and Dynamics*, 11(4), 336–342, July–August 1988.
30. Mitchell, R.L. and Walker, J.F., Recursive methods for computing detection probabilities, *IEEE Transactions on Aerospace and Electronic Systems*, AES-7(4), 671–676, July 1971.
31. Shnidman, D.A., Radar detection probabilities and their calculation, *IEEE Transactions on Aerospace and Electronic Systems*, AES-31(3), 928–950, July 1995.
32. Sandhu, G.S., A real time statistical radar target model, *IEEE Transactions on Aerospace and Electronic Systems*, AES-21(4), 490–507, July 1985.
33. Huynen, J.R., McNolty, F., and Hansen, E., Component distributions for fluctuating targets, *IEEE Transactions on Aerospace and Electronic Systems*, AES-11(6), 1316–1331, November 1975.
34. Nesline, W. and Zarchan, P., Radome induced miss distance in aerodynamically controlled homing missiles, in *AIAA Guidance Navigation and Control Conference Proceedings*, Paper No. 84-1845, Seattle, WA, 1984, pp. 99–115.
35. Borden, B., *What Is the Radar Tracking "Glint" Problem and Can It Be Solved*, Naval Air Warfare Center Weapons Division, China Lake, CA, NAWCWPNS TP 8125, May 1993.
36. Gordon, N. and Whitby, A., Bayesian approach to guidance in the presence of glint, *AIAA Journal of Guidance, Control and Dynamics*, 22(3), 478–485, May–June 1999.
37. Nesline, F.W. and Zarchan, P., Miss distance dynamics in homing missiles, *AIAA Guidance and Control Conference Proceedings*, Paper No. 84-1844, Snowmass, CO, pp. 84–98, 1984.
38. Garnell, P. and East, D.J., *Guided Weapon Control Systems*, Pergamon Press, Oxford, UK, 1977.
39. Trapp, R.L., Pulse Doppler radar characteristics, limitations and trends, The JHU/APL, Howard County, MD, FS-84-167, October 1984.
40. Farrell, J.L. and Taylor, R.L., Doppler radar clutter, *IEEE Transactions on Aerospace and navigational Electronics*, ANE-11(3), 162–172, September 1964.
41. Skolnik, M.I., *Radar Handbook*, McGraw-Hill, New York, 1970.
42. Marcum, J.I. and Swerling, P., Studies of target detection by pulsed radars, *IRE Transactions on Information Theory*, IT-6(2), 59–308, April 1960.
43. Swerling, P., Recent developments in target models for radar detection analysis, in presented at *the AGARD Avionics Technical Symposium on Advanced Radar Systems*, Istanbul, Turkey, May 25–29, 1970.

44. Xin, M. et al., Nonlinear bank-to-turn/skid-to-turn missile outer-loop/inner-loop autopilot design with θ-D techniques, in *AIAA Guidance Navigation and Control Conference Proceedings*, Austin TX, August 11–13, 2003.
45. Wassom, S.R. and Faupell, L.C., Integrated aerofin/thrust vector control for tactical missiles, *AIAA Journal of Propulsion*, 7(3), 374–381, May–June 1991.
46. Roskam, J., *Airplane Flight Dynamics and Automatic Flight Controls*, Part II, Roskam Aviation and Engineering Corporation, Ottawa, KS, 1979.
47. Nachtigal, C.L., *Instrumentation and Control, Fundamentals and Applications*, Wiley Series in Engineering Practice, John Wiley & Sons, New York, 1990.
48. Goldshine, G.D. and Lacy, G.T., High-response electromechanical control actuator, in *JPL Proceedings of the Third Aerospace Mechanisms Symposium*, Pasadena, CA, October 1, 1968, pp. 19–26.
49. Wingett, P.T. et al., Flight control actuation system, U.S. Patent 6,827,311 B2, December 7, 2004.
50. Denel Sales Brochure, Servo actuators, Broc0253.CDR, October 2007.
51. Knauber, R.N., Computer program for post-flight evaluation of the control surface response for an attitude controlled missile, NASA Contractor Report 166032, Hampton, VA, November 1982.
52. Abzug, M.L., *Computational Flight Dynamics*, AIAA Education Series, American Institute of Aeronautics and Astronautics, Reston, VA, 1998.
53. Blakelock, J.H., *Automatic Control of Aircraft and Missiles*, 2nd edn., John Wiley & Sons, New York, 1991.
54. Wise, K.A. and Broy, D.J., Agile missile dynamics and control, *AIAA Journal of Guidance, Control and Dynamics*, 21(3), 441–449, May–June 1998.
55. Wise, K.A., Comparison of six robustness tests evaluating missile autopilot robustness to uncertain aerodynamics, *AIAA Journal of Guidance, Control and Dynamics*, 15(4), 861–870, July–August 1992.
56. Wise, K.A., Missile autopilot robustness using the real multiloop stability margin, *AIAA Journal of Guidance, Control and Dynamics*, 16(2), 354–362, March–April 1993.
57. Bar-On, J.R. and Adams, R.J., Linearization of a six degree[s]-of freedom missile for autopilot analysis, *AIAA Journal of Guidance, Control and Dynamics*, 21(1), 184–187, 1998.
58. Stewart, E.C. and Smith, G.L., *The Synthesis of Optimum Homing Missile Guidance Systems with Statistical Inputs*, NASA Memo 2-13-59A, Ames Research Center, Moffett Field, CA, April 1959.
59. Stallard, D.V., *Classical and Modern Guidance of Homing Interceptor Missiles*, Raytheon Company-Seminar of Department of Aeronautics and Astronautics, Massachusetts Institute of Technology, Cambridge, MA, April 1968.
60. Calamia, R.J., Franleschetti, G., and Givli, D., Radar tracking of low altitude targets, *IEEE, Transactions Aerospace and Electronic Systems*, AES-10(4), 543–549, 1974.
61. Nesline, F.W., Missile guidance for low-altitude air defense, *AIAA Journal of Guidance, Control, and Dynamics*, 2(4), 283–289, July–August 1979.
62. Bucco, D. and Hu, Y., A comparative assessment of various multipath models for use in missile simulation studies, in *AIAA Modeling and Simulation Technologies Conference and Exhibition*, AIAA-2000-4286, Denver, CO, August 14–17, 2000.
63. Yanushevsky, R., *Modern Missile Guidance*, CRC Press, Taylor & Francis Group, Boca Raton, FL, 2008.

64. Crassidis, J.L. and Junkins, J.L., *Optimal Estimation of Dynamic Systems*, Applied Mathematics and Nonlinear Science Series, Chapman & Hall/CRC Press, Boca Raton, FL, 2004.
65. Gurfil, P., Jodorkosky, M., and Guelman, M., Finite time stability approach to proportional navigation systems analysis, *AIAA Journal of Guidance, Control and Dynamics*, 21(6), 853–861, November–December 1998.
66. Guelman, M., A qualitative study of proportional navigation, *IEEE Transactions on Aerospace and Electronic Systems*, AES-7(4), 637–643, July 1971.
67. Guelman, M., The closed form solution of true proportional navigation, *IEEE Transactions on Aerospace and Electronic Systems*, AES-12(4), 472–482, July 1976.
68. Adler, F.P., Missile guidance by three dimensional proportional navigation, *Journal of Applied Physics*, 27(5), 500–507, May 1956.
69. Cochran, J.E., No, T.S., and Thaxton, D.G., Analytical solutions to a guidance problem, *AIAA Journal of Guidance, Control and Dynamics*, 14(1), 117–122, January–February 1991.
70. Song, S., A Lyapunov-like approach to performance analysis of 3-dimensional pure PNG laws, *IEEE Transactions on Aerospace and Electronic Systems*, 30(1), 238–247, January 1994.
71. Yang, C.D. and Yang, C.C., Analytical solution of generalized three-dimensional proportional navigation, *AIAA Journal of Guidance, Control and Dynamics*, 19(3), 569–577, May–June 1996.
72. Yuan, P.J. and Chern, J.S., Solutions of true proportional navigation for maneuvering and non-maneuvering targets, *AIAA Journal of Guidance, Control and Dynamics*, 15(1), 268–271, January–February 1992.
73. Yueh, W.R., Augmented proportional navigation in third-order predictive scheme, U.S. Patent 4,502,650, March 5, 1985.
74. Biggers, E.L., Air-to-air missile guidance and control, Unpublished Technical Memorandum, July 10, 1969.
75. Yanushevsky, R.T. and Boord, W.J., New approach to guidance law design, *AIAA Journal of Guidance, Control and Dynamics*, 28(1), 162–166, January–February 2005.
76. Yanushevsky, R.T., Concerning Lyapunov-based guidance, *AIAA Journal of Guidance, Control and Dynamics*, 29(2), 509–511, March–April 2006.
77. Blake, L., Radar Range-Performance Analysis, Munro Publishing Company, Silver Spring, MD, p.39, 1991.

7
Preliminary Systems Design Trade Analysis

In this chapter, an air and missile defense (AMD) system preliminary design is evaluated against the requirements using methods described in the previous chapters and McEachron [1]. Figure 7.1 shows a graphical depiction of how the AMD battlespace unfolds from first detection to intercept assuming three interceptor variants having three different minimum ranges. Specifically, the engagement phases are as follows: range at first detection (R/detect), range at transition to a hostile track (R/firm track), range at interceptor away (R/interceptor away), range at first and subsequent nonminimum range intercepts, and minimum intercept range. This division is in fact general in that it is identical for air and ballistic targets. The timeline as presented assumes that either a subsonic or a supersonic target is being engaged at a low altitude and that the first detection occurs approximately at the horizon. At the far right, the defended area, the *keep-out zone*, is where the AMD system is presumed to be collocated within the indicated volume.

To complete a preliminary design level *battlespace analysis*, we first consider first defining battlespace depth of fire (DOF) or firepower followed by an engagement analysis. Defining the DOF requires determining for each AMD preliminary design configuration where, how many, and which interceptor variants can reach the target sets. The engagement analysis will tell us which interceptors and variants can successfully engage the targets and how many it will take to achieve the system P_k requirement. The target set is defined by speed, altitude, signature, and other environmental considerations for the battlespace evaluation. The engagement analysis requires the addition of any target defense penetration features that are uniquely intended to defeat the interceptor such as evasive maneuver [2].

For a notional AMD system preliminary design, battlespace depth-of-fire performance is examined for three missile interceptors having three minimum intercept ranges, three radar variants, and three propagation factor environments. All battlespace depth-of-fire results are for the target parametric conditions of radially inbound speeds of Mach 1–3.5 at constant altitudes of 5, 10, and 50 m [1]. In addition, the target has a nonfluctuating, Swerling 0, radar cross section that varies parametrically from 0 to −25 dBsm [1]. All results are based on engagement timeline analysis of nonmaneuvering targets. The defensive system radar and missile interceptor launcher are colocated.

Radar parameters for the depth-of-fire results correspond to the active array architecture with the exception of pulse width, which is 5 μs, supporting a radar minimum range of 0.75 km against low-altitude targets.

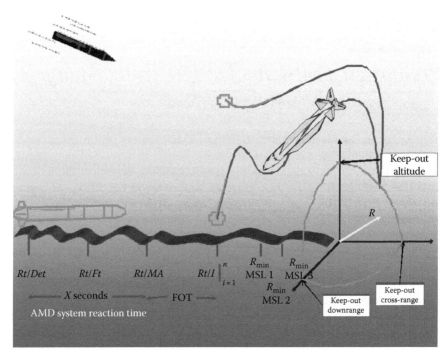

FIGURE 7.1
AMD system battlespace and engagement analysis breakdown.

This range is assumed inside the minimum intercept range for all interceptors considered. The radar detection ranges are based on single-pulse performance. It is assumed that clutter is canceled to 10 dB below thermal noise leaving the propagation factor to reduce radar sensitivity. A propagation factor of 40 dB corresponds to stressing overland engagements, while a propagation factor of 0 dB corresponds to an ideal performance reference case. Propagation factors are modeled as a constant average value at all target ranges and altitudes. Other important parameters (see McEachron [1]) used in battlespace analysis are an antenna height of 18 m, an optimum best-case combat system reaction time of 10 and 20 seconds when a kill assessment, look, is required.

The constant-altitude, inbound targets cross the radar horizon at the ranges summarized in Table 7.1. These ranges are based on a 1.62 earth model that is a good approximation for propagation conditions over a sea environment.

The radar horizon crossing ranges define the furthest range at which the target can be detected if the radar has adequate sensitivity and any clutter returns have been suppressed to levels at least 10 dB below thermal noise. Beyond these ranges, the targets are obscured from detection by the physical radar horizon.

Preliminary Systems Design Trade Analysis

TABLE 7.1

Target Radar Horizon Range Summary

Target Altitude (m)	Radar Horizon Crossing Range (km)
5	28
10	32
50	49

Figure 7.2 shows that the three targets can be detected at the radar horizon for radar cross sections as low as −25 dBsm for the case of baseline radar sensitivity and a 0 dB propagation factor. The plot legend identifies the target altitudes. The detection range performance is for the case of a 0.9 probability of detection with 1×10^{-6} probability of false alarm. When the propagation factor increases to 20 dB, none of the targets are detectable at the radar horizon for radar cross sections below approximately −17 dBsm as shown in Figure 7.3. Sidelobe jamming can potentially reduce the target detection ranges shown in Figures 7.2 and 7.3.

The flyout performance for notional short-range (SR), medium-range (MR), and long-range (LR) air defense interceptors (ADIs) are summarized in Figure 7.4. Notional interceptor flyout times are plotted as a function of

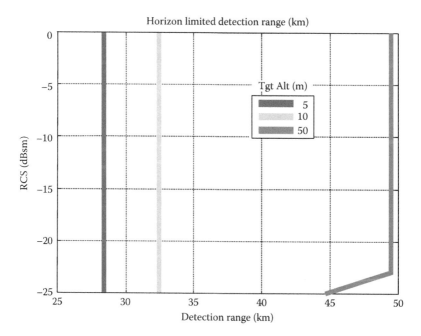

FIGURE 7.2
Radar detection ranges—baseline radar sensitivity and 0 dB propagation factor.

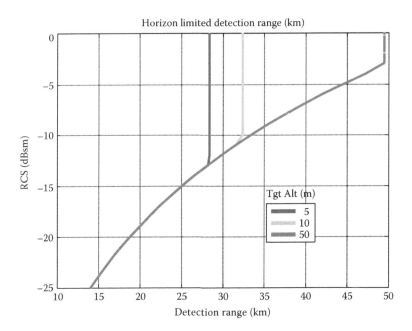

FIGURE 7.3
Radar detection ranges—baseline radar sensitivity and 20 dB propagation factor.

the range of the target when intercepted. Average velocities for the notional interceptors were estimated based on the time to fly to their maximum range. Interceptor average velocities and notional minimum intercept ranges are summarized in Table 7.2.

The required target detection ranges for the SR, MR, and LR interceptor variants are shown in Figures 7.5 through 7.7, respectively, for the four firing doctrines considered include: Shoot (S), Shoot–Shoot (SS), Shoot–Shoot–Shoot (SSS), and Shoot–Look–Shoot (SLS). For higher speed targets, the required detection ranges can be beyond the horizon-limited radar detection ranges. The required detection ranges are interceptor dependent and include the system reaction, interceptor salvo, and look times as appropriate for the firing doctrine.

7.1 Battlespace Performance Summary

A battlespace performance chart as defined in this publication is a composite of some large but finite number of AMD system battlespace evaluation elements against singular, nonparametric, conditions. A notional battlespace

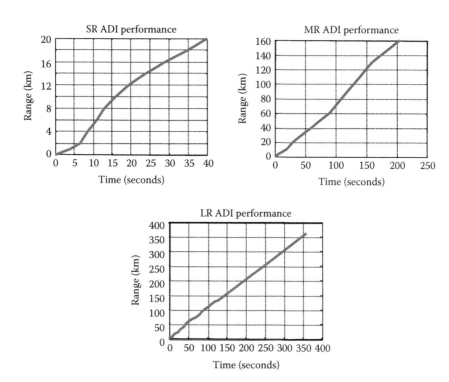

FIGURE 7.4
Notional interceptor flyout times.

chart element is shown in Figure 7.8. Figure 7.8 is a graph of cumulative probability of track initiation (P_{Dc}) for the AMD system plotted against the target range to the defended asset. The radar P_{Dc} threshold is set to 90% (horizontal line) and the radar performance P_{Dc} is a function of target signature and altitude, AMD physical configuration, the environment, and radar settings. Vertical lines are a function of interceptor flyout time, target trajectory and Mach, weapon system doctrine, and time constant. The vertical line labeled as S indicates that a shoot opportunity is possible in this example.

To create a complete battlespace chart requires spanning the target set Mach and signature spread, altitude, and environmental requirements.

TABLE 7.2

ADI Average Velocities: Minimum Intercept Ranges

AD Interceptor	Average Velocity (Mach #)	Minimum Intercept Range (km)
SR	1.5	2
MR	2.4	6
LR	2.9	10

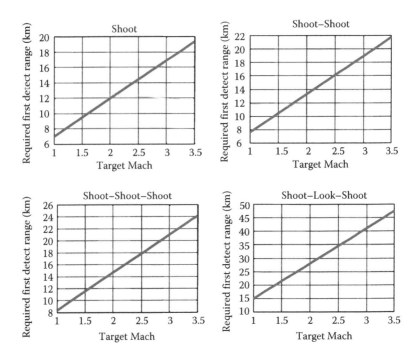

FIGURE 7.5
Required detection ranges—SR ADI.

The overall performance of the AMD system for the three missile ADIs, three radar variants, and three propagation factor environments is summarized in Table 7.3. The target conditions are radially inbound speeds of Mach 1–3.5 at constant altitudes of 5, 10, and 50 m for nonfluctuating radar cross sections of 0 to −25 dBsm. The radar performance is best balanced to the ADI for the case of 12 dB additional radar sensitivity and a 20 dB propagation factor. Depth-of-fire performance is similar for the case of baseline radar sensitivity in a 0 dB propagation factor environment. Depth-of-fire performance is somewhat degraded for both the case of 12 dB additional radar sensitivity in a 40 dB propagation factor (overland engagement) environment and the case of baseline radar sensitivity and a 20 dB propagation factor.

The battlespace depth-of-fire results' summary plots for baseline radar sensitivity with 0 dB propagation factor case are shown in Figures 7.9 through 7.17. A 0 dB propagation factor is representative of a very benign or best-case environment. The vertical boundaries between the different firing doctrines (e.g., Shoot and Shoot–Shoot) indicate that the depth-of-fire performance is the same for targets with radar cross sections between 0 and −25 dBsm. This indicates that the radar has adequate sensitivity to detect the target when it crosses the physical radar horizon. These boundaries occur at different target speeds based on the performance variations between the SR,

Preliminary Systems Design Trade Analysis

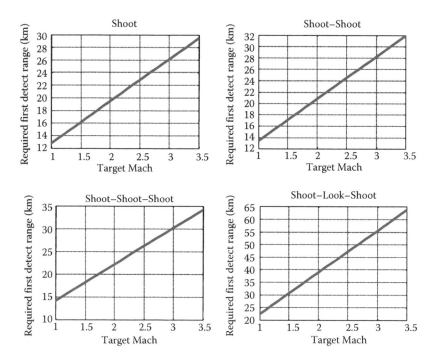

FIGURE 7.6
Required detection ranges—MR ADI.

MR, and LR ADIs. Comparing Figures 7.9 through 7.11, which summarize performance against the 5 m altitude target, we can see this variation in performance with interceptor type. The boundary between Shoot–Look–Shoot and Shoot–Shoot–Shoot is approximately Mach 2.1 for the SR interceptor, Mach 1.3 for the MR interceptor, and Mach 1.15 for the LR interceptor. This transition decreases in velocity with increased interceptor range capability. Increasing the minimum intercept range with interceptor velocity (see Table 7.2) is responsible for this trend. For the 10 m altitude target (see Figures 7.12 through 7.14), the maximum speed target that can be engaged with the Shoot–Look–Shoot doctrine improves for all three interceptors because the radar horizon has increased and the radar has adequate sensitivity to detect the target at or near the radar horizon (see Figure 7.2). A similar trend can be observed for the 50 m target in Figures 7.15 through 7.17.

Similar depth-of-fire performance is achieved with 12 dB increased sensitivity radar in a 20 dB propagation factor environment as shown in Figures 7.18 through 7.20 for the case of the LR interceptor. Depth-of-fire performance is somewhat degraded for the higher target velocities and lowest radar cross sections. For these cases, the target is not detected immediately when it crosses the radar horizon due to insufficient radar sensitivity (see Figure 7.3).

178   Air and Missile Defense Systems Engineering

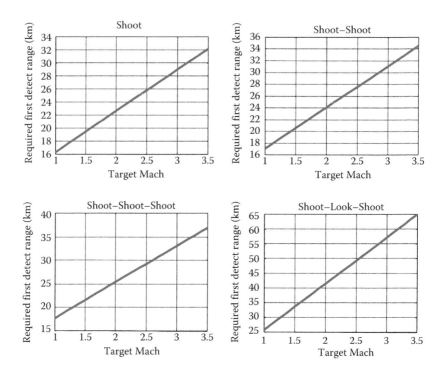

FIGURE 7.7
Required detection ranges—LR ADI.

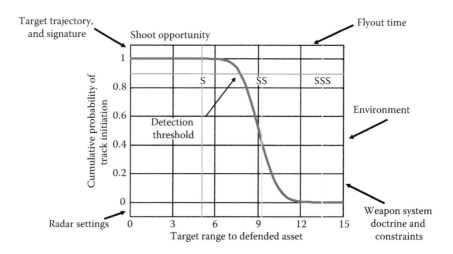

FIGURE 7.8
Notional battlespace chart element—one target (signature, Mach).

Preliminary Systems Design Trade Analysis

TABLE 7.3

Overall AMD System Performance Summary

Radar Sensitivity	Propagation Factor (dB)	System Performance Summary
Baseline	0	Radar and missile performance is balanced for all targets.
Baseline +12 dB	20	Radar and missile performance is balanced for most targets.
Baseline +12 dB	40	Radar and missile performance is balanced for some targets.
Baseline	20	Radar and missile performance is balanced for some targets.
Baseline −12 dB	40	Radar sensitivity is *not adequate* to support missile performance.
Baseline −12 dB	20	Radar sensitivity is *not adequate* to support missile performance.
Baseline −12 dB	0	Radar sensitivity is *not adequate* to support missile performance.
Baseline	40	Radar sensitivity is *not adequate* to support missile performance.
Baseline +12 dB	0	Missile performance is *not adequate* to take advantage of radar sensitivity.

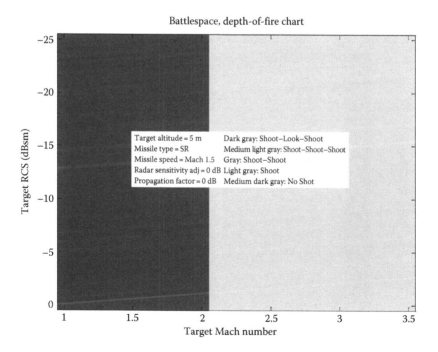

FIGURE 7.9
SR interceptor—5 m target, baseline radar sensitivity—0 dB propagation factor.

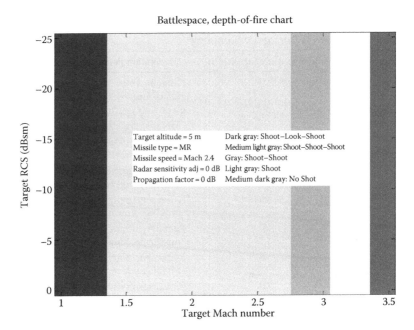

FIGURE 7.10
MR interceptor—5 m target, baseline radar sensitivity—0 dB propagation factor.

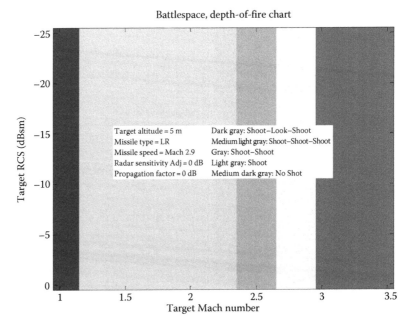

FIGURE 7.11
LR interceptor—5 m target, baseline radar sensitivity—0 dB propagation factor.

Preliminary Systems Design Trade Analysis

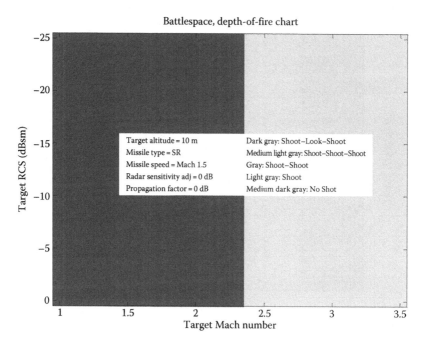

FIGURE 7.12
SR interceptor—10 m target, baseline radar sensitivity—0 dB propagation factor.

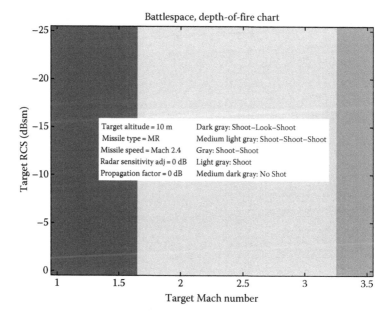

FIGURE 7.13
MR interceptor—10 m target, baseline radar sensitivity—0 dB propagation factor.

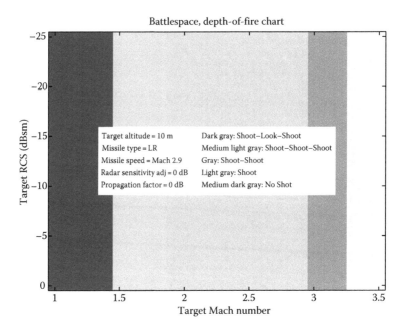

FIGURE 7.14
LR interceptor—10 m target, baseline radar sensitivity—0 dB propagation factor.

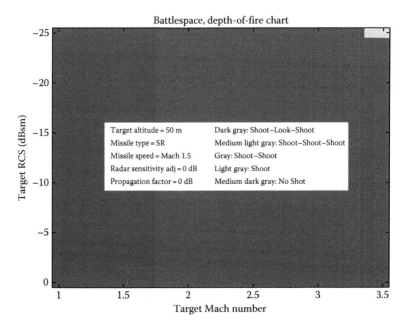

FIGURE 7.15
SR interceptor—50 m target, baseline radar sensitivity—0 dB propagation factor.

Preliminary Systems Design Trade Analysis

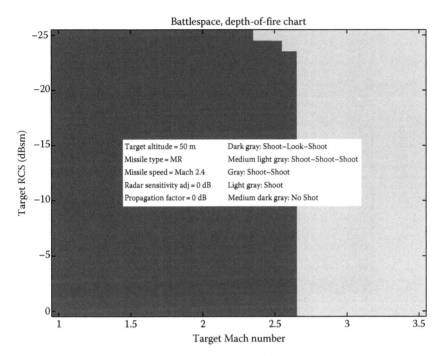

FIGURE 7.16
MR interceptor—50 m target, baseline radar sensitivity—0 dB propagation factor.

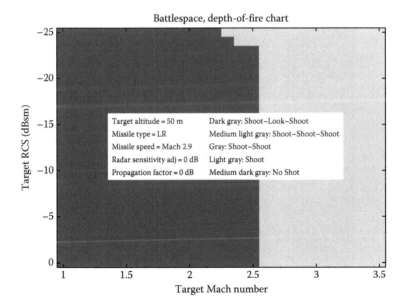

FIGURE 7.17
LR interceptor—50 m target, baseline radar sensitivity—0 dB propagation factor.

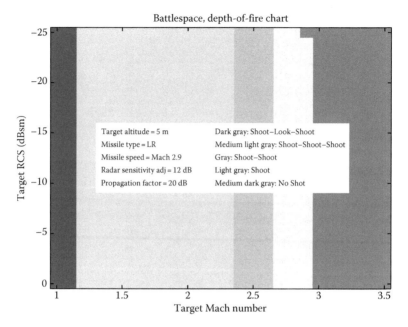

FIGURE 7.18
LR interceptor—5 m target, 12 dB increased radar sensitivity—20 dB propagation factor.

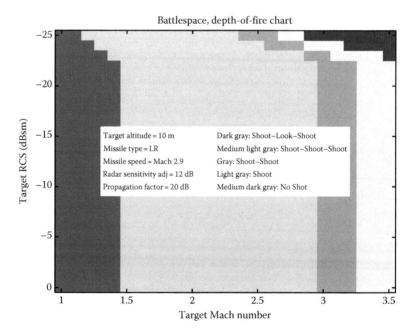

FIGURE 7.19
LR interceptor—10 m target, 12 dB increased radar sensitivity—20 dB propagation factor.

Preliminary Systems Design Trade Analysis

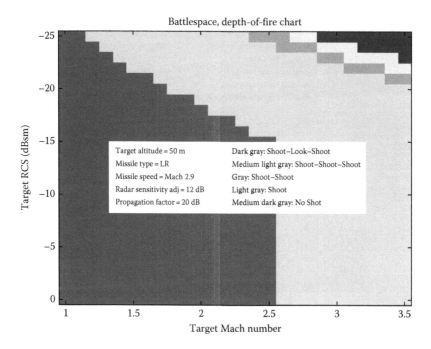

FIGURE 7.20
LR interceptor—50 m target, 12 dB increased radar sensitivity—20 dB propagation factor.

Overland scenarios are one of the more stressful environments for engaging low-altitude targets. Propagation factors can be very high. Propagation factors of 40 dB or higher are not uncommon in overland scenarios. Depth-of-fire performance for the case of 12 dB increased radar sensitivity with a 40 dB propagation factor is summarized in Figures 7.21 through 7.29. In this environment, the detection capabilities of the radar with 12 dB increased radar sensitivity are not adequate to detect some of the lower radar cross-section targets when they cross the radar horizon. This reduces the depth-of-fire performance for lower radar cross-section targets. Increased radar sensitivity and/or increased interceptor speed can be used to improve performance against the lower radar cross-section threats in an overland engagement environment.

In summary, the maximum target speed and minimum RCS capability, for the case of 12 dB increased radar sensitivity with a 40 dB propagation factor, are summarized in Table 7.4 for the Shoot–Look–Shoot firing doctrine. Table 7.4 breaks out the Shoot–Look–Shoot firing doctrine depth-of-fire performance by interceptor type and target altitude. In general, the short-range (SR) interceptor provides the best performance for low-altitude targets since the available engagement timeline is limited by the physical radar horizon and/or high propagation factors. In addition, the SR interceptor has the lowest minimum intercept range.

186   *Air and Missile Defense Systems Engineering*

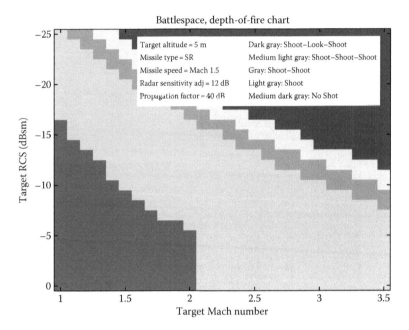

FIGURE 7.21
SR interceptor—5 m target, 12 dB increased radar sensitivity—40 dB propagation factor.

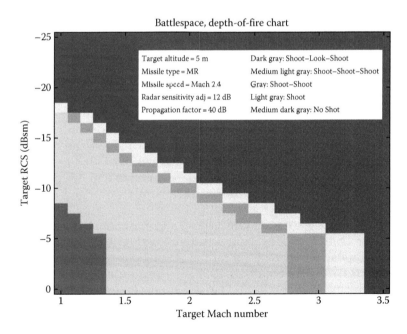

FIGURE 7.22
MR interceptor—5 m target, 12 dB increased radar sensitivity—40 dB propagation factor.

Preliminary Systems Design Trade Analysis

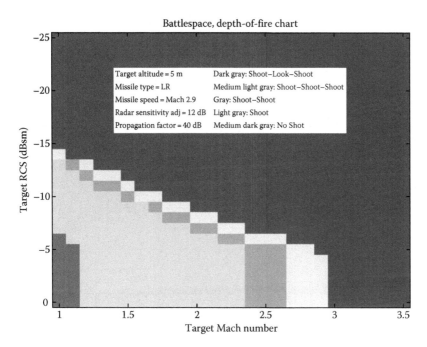

FIGURE 7.23
LR interceptor—5 m target, 12 dB increased radar sensitivity—40 dB propagation factor.

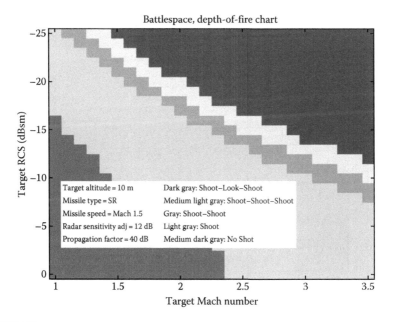

FIGURE 7.24
SR interceptor—10 m target, 12 dB increased radar sensitivity—40 dB propagation factor.

188 *Air and Missile Defense Systems Engineering*

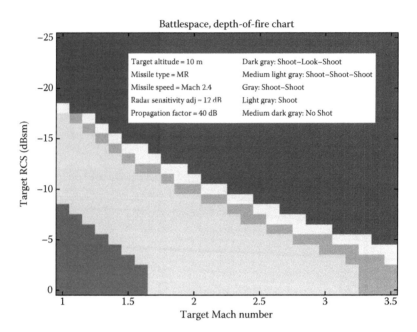

FIGURE 7.25
MR interceptor—10 m target, 12 dB increased radar sensitivity—40 dB propagation factor.

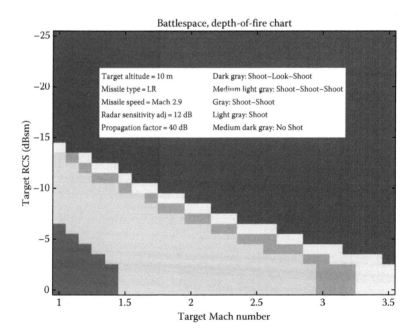

FIGURE 7.26
LR interceptor—10 m target, 12 dB increased radar sensitivity—40 dB propagation factor.

Preliminary Systems Design Trade Analysis

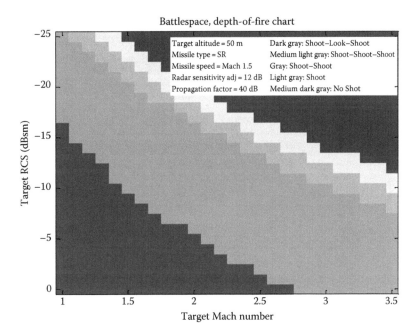

FIGURE 7.27
SR interceptor—50 m target, 12 dB increased radar sensitivity—40 dB propagation factor.

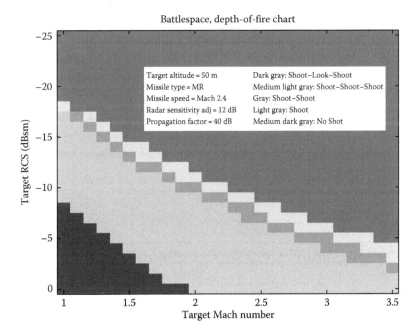

FIGURE 7.28
MR interceptor—50 m target, 12 dB increased radar sensitivity—40 dB propagation factor.

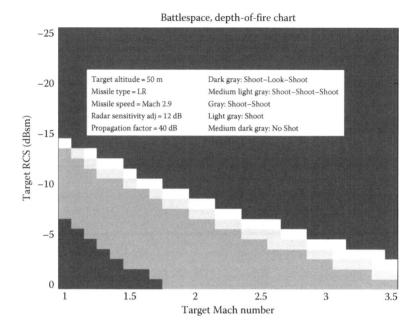

FIGURE 7.29
LR interceptor—50 m target, 12 dB increased radar sensitivity—40 dB propagation factor.

As the target height increases, the radar can detect the target sooner if it has adequate sensitivity. Earlier detection allows higher-speed targets to be engaged with a given interceptor. Table 7.4 shows how a target engineer can potentially defeat a weapon system by increasing target speed and/or lowering target RCS. The effectiveness of a weapon system improves with a layered defense approach. This allows the most effective weapon(s) to be selected based on the range and speed of the target at the time of radar detection.

After developing the answers to the questions how many and which interceptors can be delivered to the targets in question, it is necessary to answer

TABLE 7.4

Shoot–Look–Shoot Firing Doctrine Effectiveness Summary: 12 dB Increased Radar Sensitivity, 40 dB Propagation Factor

Engagement Capability (Max Mach #/Min RCS in dBsm)			
	Target Altitude (m)		
Missile Type	5	10	50
SR	Mach 2.1/−5 dBsm	Mach 2.3/−3 dBsm	Mach 2.8/0 dBsm
MR	Mach 1.3/−5 dBsm	Mach 1.7/−3 dBsm	Mach 1.9/0 dBsm
LR	Mach 1.1/−5 dBsm	Mach 1.4/−3 dBsm	Mach 1.7/0 dBsm

the question of single-shot probability of kill (P_{ssk}). Delivering an interceptor to a target does not necessarily mean that the interceptor can hit the target or achieve a sufficiently small miss distance against the target to cause enough damage to cause a mission kill. The interceptor must possess enough energy and a sufficiently small maneuver time constant to achieve adequately small miss distances to complete the engagement successfully. To conduct this analysis, a detailed end game, six-degree-of-freedom (6DOF) Monte Carlo miss distance simulation, is eventually required [3]. Within the preliminary design phase of development, it is reasonable and appropriate to conduct this analysis with a planar Monte Carlo–based terminal homing interceptor performance simulation, and as iterative passes provide more detailed definition of the design, it is possible to move to a true high-fidelity 6DOF. The planar simulation will include modeled seeker range-dependent and range-independent noise sources, radome boresight error, and possibly other Monte Carlo variables that will impact miss distance statistics.

Once the interceptor variants and the number of variants the system can deliver to the target is determined using the battlespace DOF analysis process, each of these interceptors is studied with planar simulations to assess miss distance performance. Figure 7.30 presents some of the results using a planar

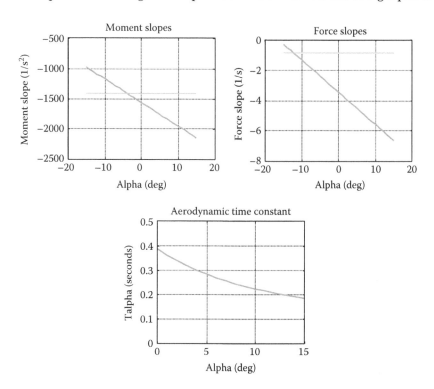

FIGURE 7.30
Interceptor angle-of-attack-dependent linearized airframe characteristics.

TABLE 7.5
Interceptor Zero Angle-of-Attack Linearized Airframe Characteristics

Z_α (1/s)	M_α (1/s²)	Z_δ (1/s)	M_δ (1/s²)
−3.5	−1550	−0.85	−1415

$\omega_z = 65.3947$, $VC = 1219$, $VM = 900$
$\omega_{af} = 39.3700$
$\zeta_{af} = 0.65$
$\eta_{limit} = 10$ g's

Monte Carlo homing analysis tool. Figure 7.31 shows Monte Carlo miss distance results for one specific engagement and 250 runs. For this example, the engagement conditions included a 3-g target step maneuver, 0.001 rad of angle noise, and a −0.01 rad/rad radome boresight error slope. The interceptors' zero degree angle-of-attack airframe characteristics are given in Table 7.5.

Monte Carlo results for a 10- and 9.5-second terminal homing time (THT) are shown in Figure 7.31 assuming that the interceptor is using true proportional navigation having a navigation constant of three and the measured states are estimated using a classic three-state Kalman filter. The plots on the right are histograms of miss distance probabilities. This is only an example; an actual

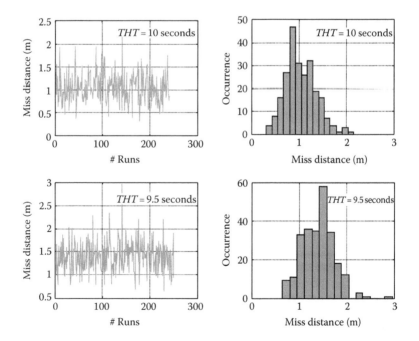

FIGURE 7.31
Example Monte Carlo engagement results.

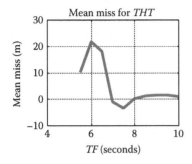
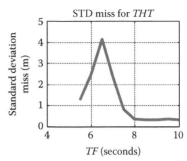

FIGURE 7.32
Monte Carlo engagement results summary.

simulation may require more than 250 Monte Carlo runs to arrive at statistically meaningful results.

A summary chart can be constructed as a function of homing time to display the mean and standard deviation of the miss distance results. The example statistical miss distance set is provided in Figure 7.32 for homing times between 10 and 5.5 seconds. The results clearly indicate that homing times less than 6.5 seconds will require a significant increase in warhead capability or a kill strategy adjustment for the notional engagement and interceptor design tested.

Table 7.6 outlines a notional engagement preliminary design study with more stressing target requirements and additional noise sources including those errors introduced at handover. The 250 Monte Carlo run results are rolled up from each case. This evaluation set focuses on interceptor g-limit requirements as a function of handover error variations. All other error sources are held constant at nominal values.

Figure 7.33 shows results for one homing time condition, 7 seconds that include a parametric evaluation of interceptors having variable acceleration limits shown on the ordinate that correspond to Table 7.6. Two additional cases were added to include interceptor g-limits of 55 and 60 g's.

Figure 7.33 shows the lethality miss distance threshold requirement to produce a 50% P_{ssk}, which would be based on a lethality strategy that would have to be a flow-down requirement. According to this design study, the lateral acceleration limit of the interceptor as modeled would need to exceed 45 g's assuming that 7 seconds is the minimal acceptable homing time. At this point, the preliminary design would proceed assuming that the airframe is capable of achieving this g-limit at the ranges necessary to satisfy the engagement boundary requirements.

A follow-on analysis shown in Figure 7.34 indicates that achieving a lethality ratio of one would require three interceptors on target to achieve the flow-down $0.9P_k$ requirement. A lethality ratio between 1 and 0.5 would be required to reduce the number of interceptors to 2. This would require a 50-g lateral acceleration capability interceptor according to Figure 7.33.

TABLE 7.6
Notional Engagement Preliminary Design Study

Case	Lateral g Limit	Handover Error HE (deg)	Handover Error CR (m)	$T\alpha$	Angle Noise (rad)	Target Maneuver (g's)	PRONAV N'	RBS (deg/deg)
1	50	10	500	0.25	0.001	15	3	−0.01
2	45	10	500	0.25	0.001	15	3	−0.01
3	40	10	500	0.25	0.001	15	3	−0.01
4	35	10	500	0.25	0.001	15	3	−0.01
5	50	5	500	0.25	0.001	15	3	−0.01
6	45	5	500	0.25	0.001	15	3	−0.01
7	40	5	500	0.25	0.001	15	3	−0.01
8	35	5	500	0.25	0.001	15	3	−0.01
9	50	0	500	0.25	0.001	15	3	−0.01
10	45	0	500	0.25	0.001	15	3	−0.01
11	40	0	500	0.25	0.001	15	3	−0.01
12	35	0	500	0.25	0.001	15	3	−0.01
13	50	10	250	0.25	0.001	15	3	−0.01
14	45	10	250	0.25	0.001	15	3	−0.01
15	40	10	250	0.25	0.001	15	3	−0.01
16	35	10	250	0.25	0.001	15	3	−0.01
17	50	5	250	0.25	0.001	15	3	−0.01
18	45	5	250	0.25	0.001	15	3	−0.01
19	40	5	250	0.25	0.001	15	3	−0.01
20	35	5	250	0.25	0.001	15	3	−0.01
21	50	0	250	0.25	0.001	15	3	−0.01
22	45	0	250	0.25	0.001	15	3	−0.01
23	40	0	250	0.25	0.001	15	3	−0.01
24	35	0	250	0.25	0.001	15	3	−0.01
25	50	10	0	0.25	0.001	15	3	−0.01
26	45	10	0	0.25	0.001	15	3	−0.01
27	40	10	0	0.25	0.001	15	3	−0.01
28	35	10	0	0.25	0.001	15	3	−0.01
29	50	5	0	0.25	0.001	15	3	−0.01
30	45	5	0	0.25	0.001	15	3	−0.01
31	40	5	0	0.25	0.001	15	3	−0.01
32	35	5	0	0.25	0.001	15	3	−0.01
33	50	0	0	0.25	0.001	15	3	−0.01
34	45	0	0	0.25	0.001	15	3	−0.01
35	40	0	0	0.25	0.001	15	3	−0.01
36	35	0	0	0.25	0.001	15	3	−0.01

Preliminary Systems Design Trade Analysis

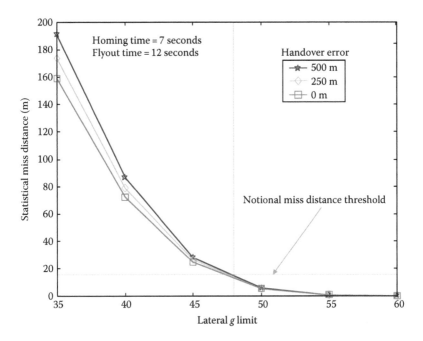

FIGURE 7.33
Monte Carlo interceptor results for 7 second homing time.

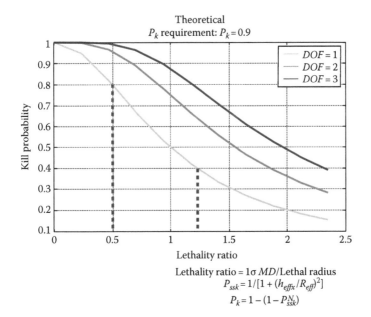

FIGURE 7.34
Theoretical kill probability as a function of DOF.

A preliminary design selection can be made once this analysis is completed for all interceptor design options against each target set and under the engagement considerations of interest. If this selection is to be based on, for example, target evasive maneuver level, then a selection chart like the one shown in Figure 7.35 would be developed. Several of these charts would be

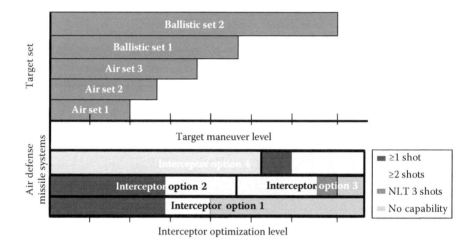

FIGURE 7.35
Interceptor evaluation map.

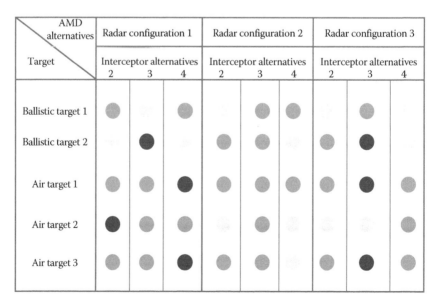

FIGURE 7.36
AMD down selection stoplight map.

required to examine interceptor design options where various target defense penetration features would be independently and in combination chosen as the independent variable.

Once the battlespace, DOF, and engagement analysis is completed, then a balanced set of AMD design options would be available either for moving into another iteration loop for preliminary design improvements or for moving to CDR. A traditional stoplight chart would be a mechanism to compile a massive amount of design and performance evaluation results into a succinct set of alternatives. Figure 7.36 shows an example of AMD down selection stoplight chart. Light gray corresponds to cases where requirements are partially met, gray corresponds to cases where requirements are met, and dark gray corresponds to cases where requirements are not met.

References

1. McEachron, J.F., Subsonic and supersonic antiship missiles: An effectiveness and utility comparison, *Naval Engineers Journal*, 109, 57–73, January 1997.
2. Yanushevsky, R., *Modern Missile Guidance*, CRC Press, Taylor & Francis Group, Boca Raton, FL, 2008.
3. Hawley, P.A. and Blaukamp, R.A., Six degree-of-freedom simulations for guidance, navigation and control, *Johns Hopkins APL Technical Digest*, 29(1), 71–84, 2010.

8
Allocation of Performance Requirements

8.1 Allocation of Radar Performance Requirements to Subsystems

The radar element requirements are developed in a similar process to the radar system requirements. The key radar elements are the antenna, transmitter, and signal/data processor. Once the target is detected, the radar must be able to accurately determine the position and heading of the target in natural and man-made environments. The position of the target is determined from measuring the range and azimuth and elevation. Each estimate of range and angle is used to update track filters that smooth the target's position and estimate the target velocity and acceleration. The target track data are used in the engagement solution to determine the target intercept point. The accuracy of the radar estimates will determine the uncertainty volume that the interceptor seeker must search in order to acquire the target and transition into terminal homing. The process for determining the radar element requirements that support target tracking accuracy is illustrated in Figure 8.1.

The key design parameters are instantaneous bandwidth, antenna beamwidth, and target signal-to-noise ratio. These radar parameters affect the accuracy of the target range and angle estimates. The antenna architecture, signal processing, and utilization of enabling technologies optimize radar performance in natural and man-made environments. The antenna architecture is usually selected to provide low sidelobes that mitigate the effects of jamming and clutter in the sidelobes. Mainlobe clutter is removed with signal processing techniques such as moving target indicator (MTI) or pulse Doppler (PD). Key enabling technologies are digital beamforming, adaptive sidelobe cancellation, and T/R modules.

The radar instantaneous bandwidth (IBW) determines the radar range resolution. The achievable radar range resolution is as follows:

$$\text{Achievable range resolution} = 240/(\text{IBW(MHz)}) \text{ m}$$

A radar with 100 MHz of instantaneous bandwidth can achieve a range resolution of 0.24 m. The achievable radar resolution includes 2 dB of processing

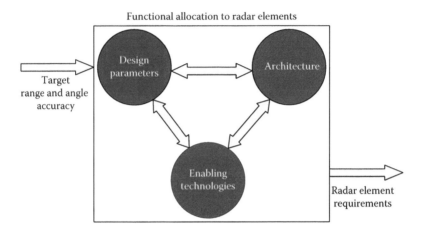

FIGURE 8.1
Radar element requirement process.

loss, which is a typical value. In order to resolve two point targets, a good rule of thumb is that they should be separated in range by at least three times the range resolution. For the case of 100 MHz IBW, the two point targets should be at least 0.72 m apart in range.

Range accuracies of 10–20 m are usually adequate to support most target engagements. Instantaneous bandwidths of 12–24 MHz are adequate to support this requirement. Higher instantaneous bandwidths are typically used for target imaging to support target ID. This is typically done in ballistic interceptor defense when it is desirable to locate the reentry vehicle in a cloud of ballistic objects including boosters and separation and thrust termination debris.

The antenna beamwidth sets the radar angular resolution. The antenna beamwidth is inversely proportional to the antenna gain. Therefore, the antenna beamwidth becomes narrower with increasing radar frequency and increasing antenna size. Radar angular resolution is typically in the order of 1/10 of a beamwidth.

Range and angle accuracy are inversely proportional to the signal-to-noise ratio and the number of track updates available to the tracking filter. Therefore, track accuracy improves over time. Figure 8.2 illustrates combinations of antenna beamwidth and number of target updates required to achieve a given track azimuth or elevation angular accuracy of 0.1° and 0.05° for a notional tracking filter design and a fixed target signal-to-noise ratio of 16 dB. An angle accuracy of 0.05° can be achieved using a simple tracking filter using a 2° beamwidth with 25 track updates at a signal-to-noise ratio of 16 dB.

Figure 8.3 shows a similar relationship versus radar frequency. Clearly for a fixed available track time (number of hits), higher frequencies and narrower beamwidths provide better tracking accuracy.

Allocation of Performance Requirements

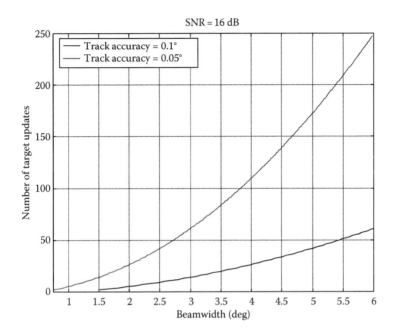

FIGURE 8.2
Number of target updates and antenna beamwidth required to support angular track accuracy.

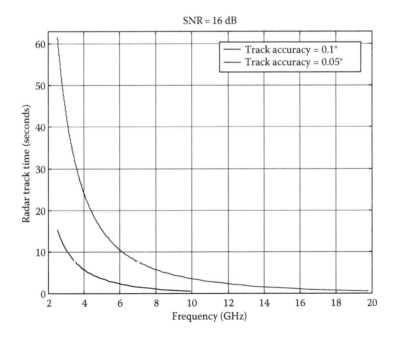

FIGURE 8.3
Radar track time and radar frequency required to support angular track accuracy.

The natural environment can affect the ability of the radar to accurately measure target position and angle. Clutter and multipath can reduce the target signal-to-noise ratio. Multipath can also degrade the angle estimation accuracy particularly for estimates of the target elevation angle.

If the main beam clutter is not canceled below thermal noise by signal processing techniques, such as MTI or PD, the clutter can reduce the effective target signal-to-noise ratio. This will reduce track accuracy and in some cases cause track drops because the target is completely masked by the clutter return. Clutter through the sidelobes must also be minimized. This is typically done with low receive pattern sidelobes and potentially some control of the transmit sidelobes in angular regions where large clutter returns are present. Low receive sidelobes also provide some immunity to sidelobe jamming. The level of clutter rejection needed in the main beam and sidelobes can be traded off in the radar design between the antenna and signal processor requirements.

Low receive channel sidelobes can be achieved with analog or digital beamforming architectures. Digital beamforming provides some additional flexibility for performing adaptive receive sidelobe control. The synthesis of low sidelobe regions in the transmit pattern can be done with the T/R modules in an active phased array. The T/R modules provide the flexibility to synthesize low sidelobe regions using the independent phase control at each element provided by the T/R modules.

Multipath is a target return from a secondary indirect path, which includes a signal reflection from the terrain between the radar and the target. Multipath levels are geometry, terrain, frequency, and polarization dependent. Multipath produces target fading that reduces signal-to-noise ratio and tracking accuracy. In some cases, the multipath is constructive and enhances the target signal-to-noise ratio by up to 6 dB if it is completely constructive.

Multipath can also corrupt the angle estimate by producing a false reflected target at the same range but at a slightly different angle. Most angle estimation techniques are degraded when multiple targets are present in the same range cell because they are unable to distinguish between the two targets. This usually results in a large estimated angle error for the actual target. As mentioned previously, multipath degrades the target elevation angle estimate. Using a narrower beamwidth reduces the effect of multipath errors. Signal processing techniques can be used to improve the target elevation angle estimate in multipath environments.

Jamming in the sidelobes can reduce the target signal-to-noise ratio and tracking accuracy. The best approach to minimizing the effects of sidelobe jamming is to use a low sidelobe receive pattern that essentially has low sidelobes everywhere. In addition, sidelobe cancellation can be employed to further reduce sidelobe jamming levels. Sidelobe cancellation can be easily implemented in digital beamforming antenna architectures. Cancellation of sidelobe jamming becomes more difficult as the bandwidth over which the jamming must be canceled increases. Therefore, cancellation of sidelobe jamming will generally be more effective for narrowband search and track modes as compared to high-bandwidth target imaging modes.

Mainbeam jamming can potentially be overcome by increasing the effective transmit power. This can be accomplished by integrating over a number of pulses to *pull* the target return out of the jamming. When the target and jammer are in the main beam but have some angular separation, techniques exist to perform main beam nulling to cancel main beam jamming. Main beam nulling can be incorporated in digital beamforming as well as analog beamforming architectures. In general, good target tracking accuracy performance is difficult to achieve in main beam jamming scenarios.

The target position accuracy is achieved primarily by the antenna and signal processor requirements. The antenna beam must be narrow enough to support angle accuracy requirements for a reasonable number of target updates. The angular accuracy becomes more important with longer target engagement ranges as the magnitude of the cross-range error associated with a fixed angle estimate increases with target range. The signal processor must be capable of canceling main beam clutter while antenna transmit and receive sidelobes must be capable of suppressing the sidelobe clutter. The combination of the antenna architecture and signal processing must also be capable of sufficiently suppressing sidelobe jamming through the combination of low receive sidelobes and adaptive sidelobe nulling.

8.2 Allocation of Interceptor Performance Requirements to Interceptor Subsystems

8.2.1 Terminal Homing and Guidance

The AMD system declares the time or range to go for transition from midcourse guidance to terminal homing for any given engagement. The interceptor terminal homing phase actually begins when the AMD system estimate of time or range to go before the predicted intercept point matches the declared time or range to go. This point in time and range to go defines the end of midcourse guidance and handover. The guidance package is where the flow-down of requirements begins.

The interceptor guidance package consists of a seeker with optics or an antenna, radome/irdome and its associated control and gimbal drive group, and a signal processor. The guidance package contains a rear reference receiver with uplink/downlink capability and its associated rear-facing antenna and a feed line. These components as a set receive RF/IR energy originating at or reflecting from target sets, receive uplink messages, reduce the received energy to target directional information, and provide guidance acceleration commands to the control and steering section to cause the interceptor airframe to fly a minimum energy course that will satisfy the

P_{ssk} requirements. In other words, the objective of the guidance and control system is to bring the interceptor within the specified miss distance in such a manner as to minimize the degrading effects of noise and error influences (e.g., clutter, spillover, or reflection) and to satisfy the kill probability for a specific kill strategy (e.g., warhead or hit to kill).

To allocate guidance accuracy requirements, the interceptor's miss distance error budget contributors versus interceptor performance measures are itemized. For example, in order to establish accurate heading error estimates, midcourse guidance, control, and navigation must be properly itemized. Heading error is a contributor to miss distance and is folded into the remaining miss distance error sources and so on. Properly accounting for heading error, established at handover, requires accurately predicting available homing time/seeker acquisition range, accurately accounting for the homing loop time constant, accurately predicting interceptor time-dependent speed and acceleration limits, and correctly predicting the guidance commands. Guidance commands will need to be provided in the presence of various noise sources to include target fade, scintillation, glint, and electronic countermeasures, and the guidance package is required to minimize the possibility of homing on unassigned targets, flight associated debris, and debris of targets previously destroyed by another interceptor.

Complications occur such as if the target begins to maneuver, then the homing loop time constant is no longer constant or linear. This requires an additional level of design and modeling to account for the flight control system time response and margins properly capturing the varying and nonlinear effects maneuver will have on the homing loop time constant.

Moreover, during the terminal homing phase, the guidance package is required to modify the trajectory and/or guidance law and/or gains for low-, medium-, or high-altitude engagements.

When the interceptor has approached the target sufficiently close to arm, it will enter the intercept phase. During this phase, the fuze or target detecting device (TDD) should detect the target and detonate the warhead assuming that a warhead kill is sought.

8.2.2 Launch and Flyout Phase

We will assume that AMD interceptors are launched from vertical launchers and have a guidance-free flight period and a preprogrammed pitch over period. During these periods, no guidance commands are generated and aerodynamic control and/or the thrust vector control (TVC) actuator systems are disabled for the time required to clear the launcher, and then control is activated to stabilize the interceptor in vertical flight for the time required for the interceptor to clear the launch area. After clearing the launch area, a controlled pitch over places the interceptor on the correct flight path. Assuming a roll attitude–stabilized interceptor after achieving the required pitch over maneuver and airframe stabilization, the airframe is rotated to the preferred steering orientation normally set prior to launch, and then the roll rate is nulled for roll attitude–stabilized systems. Once this final, nulling of roll rate, launch activity is achieved, the interceptor transitions to either midcourse guidance activation or homing if the interceptor is in a home-all-the-way mode.

During flyout, the rocket motor stack provides the kinematic performance to achieve the time to intercept and the minimum capability against maximum range and maneuvering target requirements. The propulsion system propellant, volumetric properties and hardware design, and number of stages constitute the degrees of freedom for propulsion requirements flow-down. The propulsion section design trade space contains sizing the necessary stages to boost the missile to the high-end speed and sizing the sustainer system to maintain the speed the airframe requires for affecting a successful intercept against the targets at the ranges specified in the requirements.

The flight control system (FCS) is required to stabilize the airframe and affect changes in direction or maneuverability as directed by the guidance computer. The interceptor is designed for achieving directional accelerations with an established magnitude and time requirement set either in a single aerodynamic surface or in a combined plane maneuver. The structural limitation of the interceptor guidance package or other subsystems may require the actual interceptor performance to be limited to a lesser maneuverability capability than the airframe will permit and may drive the flow-down of FCS requirements. The FCS has several components to include the controller computer that provides control and stabilization of the interceptor. The controller section receives commands from the guidance computer (sometimes when combined, these components are called an autopilot; here, we refer to this package as the G&C unit) to command the aerodynamic control fins and/or the TVC vanes to guide the interceptor on a target intercept course. The controller computer provides control authority for pitch, yaw, and roll stabilization. A servo control unit (SCU) contains the control and communications electronics necessary to convert commands from the controller to commands to the actuator system(s) and then provides controller feedback on aerodynamic control fin and/or TVC positions.

The actuator system(s) accept commands from the SCU and transforms these commands to the mechanical sources required to move the aerodynamic control fins and/or TVC vanes so that the interceptor flies the trajectory commanded by the G&C unit.

The inertial reference unit (IRU) package contains a set of gyroscopes and accelerometers used to provide relative position, motion, and accelerations as feedback to the G&C unit. The instrumentation in the IRU package must have the bandwidth to supply feedback to the G&C unit that will enhance achieving the rigorous interceptor stability and control requirements throughout the engagement space. The three IRU package design metrics are accuracy, precision, and bandwidth. The IRU package has to be designed to interface with G&C during both terminal and midcourse guidance.

The interceptor weight and balance requirements are part of the trade space to achieve flyout design requirements. Moreover, the interceptor centers of gravity and moments of inertia parameters are important metrics to the launching system and storage and handling requirements.

The interceptor body diameter and the interceptor length with and without in-line boosting systems and TVC units are part of the flyout design trade space, in that the effects of aerodynamic metrics are impacted to include drag. Moreover, the total interceptor dimensions, including aerodynamic control fins and TVC unit, if used, are required to fit within the volumetric constraints of the launching system. Fixed aerodynamic surfaces are part of the flyout design trade space that can be used to adjust the interceptor center of pressure and improve body lift over the engagement envelope. The aerodynamic control fins are sized to maintain control authority throughout the flight envelope and provide the aerodynamic gain required to stabilize and guide the interceptor to target intercept. The aerodynamic control fins may be foldable such that the interceptor will fit into the launching system.

9
Physics and Mathematics of AMD Design and Analysis

9.1 Interceptors and Flight Analysis

Modern air and missile defense (AMD) systems are required to encounter targets at ranges beyond the horizon using multiple sensors and assets for prosecuting the engagement. Search, detection, transition to track, fire control, midcourse guidance, and terminal homing, all parts of the air defense interceptor engagement, may be performed from different assets that will most likely be separated by hundreds of kilometers. These engagements are made possible by accomplishing an accurate location of each asset that is part of the engagement and passing data. The precise relative instantaneous location of assets can be accomplished using satellite-aided (e.g., GPS) navigation and targeting and accurate models of the earth geoid. The data passed onto the data links will include target positions, velocities, launch positions, target and interceptor relative positions, and velocity. Accurate flight and state representation parameters are paramount to developing firing doctrines and fire-control solutions and computing midcourse and terminal guidance commands. This section provides the fundamental mathematics of locating and translating solutions from one reference frame to another and accurately computing flight performance in a target environment.

9.1.1 A WGS-84 Oblate, Rotating Earth Model

The WGS-84 oblate, rotating earth (ORE) model [1] is likely the most widely used by the air and missile defense community and is provided in some detail here. The WGS-84 model accounts for rotation effects such as tangential and centripetal acceleration and oblateness effects like geodetic versus geocentric positioning and nonuniform gravity. Moreover, the ORE model includes rotationally induced forces and effects encountered by a missile in flight. Although the WGS-84 ORE is not a precise representation of the actual earth, because it does not take into account surface features such as mountains and higher-order shape anomalies, its ellipsoidal representation is more accurate than a spheroid with constant radius that is typically insufficient for serious AMD design and analysis.

9.1.1.1 Transformation Matrices: Coordinate Frames and Position

From Etkin [1,2], there are many frames of reference encountered in flight dynamics. For example, the earth-centered inertial and launch-centered inertial Cartesian (ECIC and LCIC, respectively) frames, the locally level vehicle (LLV) carried reference frame that is oriented to the earth surface fixed Cartesian (ESFC) frame and sometimes with the LCIC, and other various coordinate systems are used in the design and development of air and missile defense systems. The body frame is centered at the missile center of gravity, with the positive x-axis out the nose of the missile. The wind frame has the same origin as the body frame but with x-axis along the direction of the velocity vector (wind). The locally level vertical frame has the same missile center of gravity origin, but with the x-axis directed *north* toward the pole of the earth, and the y-axis directed *east* and the z-axis directed *down*, toward the center of the earth.

Transformation between any of these systems requires a multiplication by a transformation matrix that relates the orientation of each frame to one another and an addition of the distance between the origins of each reference frame being transformed. For example, to transform from LCIC to ECIC frames, a multiplication by the transformation matrix is required, along with the addition of distance between the two frame origins in Cartesian coordinates.

Figure 9.1 [1–3] provides the rotation sequence for the derivation of the transformation matrix between ECIC and LCIC given in the following equations:

$$\begin{bmatrix} X' \\ Y' \\ Z' \end{bmatrix} = B_\ell \begin{bmatrix} X \\ Y \\ Z \end{bmatrix}; \quad B_\ell = \begin{bmatrix} \cos(\ell) & \sin(\ell) & 0 \\ -\sin(\ell) & \cos(\ell) & 0 \\ 0 & 0 & 1 \end{bmatrix} \quad (9.1)$$

$$\begin{bmatrix} N \\ E \\ D \end{bmatrix} = B_\mu \begin{bmatrix} X' \\ Y' \\ Z' \end{bmatrix}; \quad B_\mu = \begin{bmatrix} -\sin(\mu) & 0 & \cos(\mu) \\ 0 & 1 & 0 \\ -\cos(\mu) & 0 & \sin(\mu) \end{bmatrix} \quad (9.2)$$

$$\begin{bmatrix} X_{az} \\ Y_{az} \\ Z_{az} \end{bmatrix} = B_\lambda \begin{bmatrix} N \\ E \\ D \end{bmatrix}; \quad B_\lambda = \begin{bmatrix} \cos(\lambda) & \sin(\lambda) & 0 \\ -\sin(\lambda) & \cos(\lambda) & 0 \\ 0 & 0 & 1 \end{bmatrix} \quad (9.3)$$

The transformation shown in Figure 9.1 only involves rotations to orient the axes in angle along the LCIC frame launch azimuth relative to the ECIC

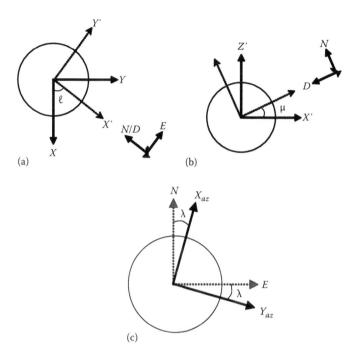

FIGURE 9.1
ECIC to LCIC transformation matrix rotation sequence: (a) First rotation: Earth top view (Equation 9.1); (b) Second rotation: Earth side view (Equation 9.2); and (c) Third rotation: Above launch point (Equation 9.3).

frame and does not involve any position translation. Thus, with this matrix, a point in space defined in the ECIC coordinate system can be defined in the LCIC frame. The first rotation (ℓ) shown in Figure 9.1 is a rotation along the earth longitudinal angle about the ECIC vertical z-axis to locate the launch frame in the easterly plane. The second rotation (μ) is a rotation about the ECIC X–Y plane in earth latitude to locate the launch frame in the northerly plane. After carrying out the two previous rotations, the resulting axis is a north–east–down (NED) system, with each axis pointing in one of those three directions. The launch frame, however, may not be aligned with the NED frame, and thus, the third rotation (λ) aligns the launch frame along the firing azimuth. The LCIC z-axis always aligns toward the center of earth. The alignment demonstrated in Figure 9.1 assumes a spherical earth model noting that there is no difference between geodetic and geocentric latitude.

Combining all three rotation matrices by multiplication, the overall ECIC to launch the transformation matrix is achieved, shown in Figure 9.2 [3], and the resulting transformation matrix is given in the following equations:

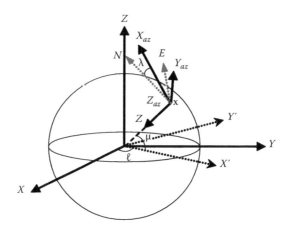

FIGURE 9.2
Spherical nonrotating earth ECIC to LCIC transformation matrix.

$$\begin{bmatrix} X_{az} \\ Y_{az} \\ Z_{az} \end{bmatrix} = B_\lambda B_\mu B_\ell \begin{bmatrix} X \\ Y \\ Z \end{bmatrix}; \quad B = B_\lambda B_\mu B_\ell; \quad B = \begin{bmatrix} b_{11} & b_{12} & b_{13} \\ b_{21} & b_{22} & b_{23} \\ b_{31} & b_{32} & b_{33} \end{bmatrix} \quad (9.4)$$

$$b_{11} = -\cos(\lambda)\sin(\mu)\cos(\ell) - \sin(\lambda)\sin(\ell)$$
$$b_{12} = -\cos(\lambda)\sin(\mu)\sin(\ell) + \sin(\lambda)\cos(\ell)$$
$$b_{13} = \cos(\lambda)\cos(\mu)$$
$$b_{21} = \sin(\lambda)\sin(\mu)\cos(\ell) - \cos(\lambda)\sin(\ell)$$
$$b_{22} = \sin(\lambda)\sin(\mu)\sin(\ell) + \cos(\lambda)\cos(\ell) \quad (9.5)$$
$$b_{23} = -\sin(\lambda)\cos(\mu)$$
$$b_{31} = -\cos(\mu)\cos(\ell)$$
$$b_{32} = -\cos(\mu)\sin(\ell)$$
$$b_{33} = -\sin(\mu)$$

With a nonrotating earth, the position of the launch frame with respect to ECIC coordinates is fixed, assuming that the launch frame does not translate during the flight of the missile. There are no rotation effects in the

Physics and Mathematics of AMD Design and Analysis

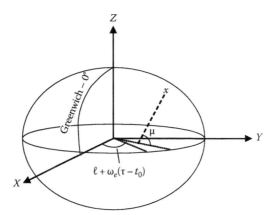

FIGURE 9.3
Time-dependent effects due to a rotating earth.

transformation from one frame to the other. Modeling a rotating earth, however, requires the position of a stationary launch frame on the surface of the earth to change with respect to the ECIC frame. The earth surface rotates with a fixed angular velocity about the axis of rotation (normally assumed to be straight through the pole). This change in position is dependent on time relative to an initial time—t_0 seconds.

At t_0, the launch frame is fixed relative to the Greenwich meridian assumed as the plane through which the ECIC x-axis passes. At time $t_0 + \tau$, the launch frame's angular position has changed as a function of the earth rotation rate vector, and the position vector change is also related to the location of the frame relative to the equator (μ).

The [B] transformation matrix will need to be modified by substituting $\ell + \omega(\tau - t_0)$ for ℓ to account for the change in position of the launch frame with time due to the rotation of the earth's surface about its center, where τ is the time from initial conditions. The effect is illustrated in Figure 9.3 [3].

The new transformation matrix will be labeled B_{RE} to denote the rotating earth transformation from ECIC. Notable is the fact that this time-dependent term only needs to be added to the ECIC-to-Launch Frame transformation matrix or similar earth surface fixed frames and any other transformations dependent on earth rotation effects. The new general model is depicted in Figure 9.4 [3]:

$$\begin{bmatrix} X_{az} \\ Y_{az} \\ Z_{az} \end{bmatrix} = B_{RE} \begin{bmatrix} X \\ Y \\ Z \end{bmatrix}; \quad B_{RE} = \begin{bmatrix} b_{11}^{RE} & b_{12}^{RE} & b_{13}^{RE} \\ b_{21}^{RE} & b_{22}^{RE} & b_{23}^{RE} \\ b_{31}^{RE} & b_{32}^{RE} & b_{33}^{RE} \end{bmatrix} \quad (9.6)$$

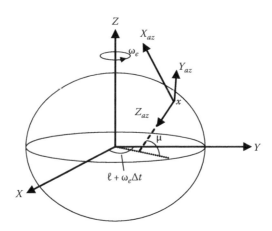

FIGURE 9.4
ECIC rotating earth to an LCIC transformation.

$$b_{11}^{RE} = -\cos(\lambda)\sin(\mu)\cos(\ell + \omega_e \cdot \Delta t) - \sin(\lambda)\sin(\ell + \omega_e \cdot \Delta t)$$

$$b_{12}^{RE} = -\cos(\lambda)\sin(\mu)\sin(\ell + \omega_e \cdot \Delta t) + \sin(\lambda)\cos(\ell + \omega_e \cdot \Delta t)$$

$$b_{13}^{RE} = \cos(\lambda)\cos(\mu)$$

$$b_{21}^{RE} = \sin(\lambda)\sin(\mu)\cos(\ell + \omega_e \cdot \Delta t) - \cos(\lambda)\sin(\ell + \omega_e \cdot \Delta t)$$

$$b_{22}^{RE} = \sin(\lambda)\sin(\mu)\sin(\ell + \omega_e \cdot \Delta t) + \cos(\lambda)\cos(\ell + \omega_e \cdot \Delta t) \quad (9.7)$$

$$b_{23}^{RE} = -\sin(\lambda)\cos(\mu)$$

$$b_{31}^{RE} = -\cos(\mu)\cos(\ell + \omega_e \cdot \Delta t)$$

$$b_{32}^{RE} = -\cos(\mu)\sin(\ell + \omega_e \cdot \Delta t)$$

$$b_{33}^{RE} = -\sin(\mu)$$

9.1.1.2 Transformation Matrices: Velocity and Acceleration

Modeling a rotating earth requires the introduction of a time-dependent coordinate frame component when developing transformations having to do with either velocity or acceleration [1,2]. Applying the chain rule to the transformation of a position vector p_a in one frame of reference to another p_b given in Equation 9.8 to obtain the velocity transformation given in Equation 9.9 reveals that it is necessary to determine if the transformation matrix is time dependent:

$$p_b = L_b^a \cdot p_a \tag{9.8}$$

$$\dot{p}_b = L_{ba}\dot{p}_a + \dot{L}_{ba}p_a \tag{9.9}$$

Notice that when the derivative of the transformation matrix exists, it is necessary to transform either velocity or acceleration vectors.

Another effect of a rotating earth that will require transformation is the additional centripetal and tangential accelerations [1]. Tangential acceleration exists since there is a change in speed of the interceptor due to the rotation of the earth, and centripetal acceleration is created from the change in direction of the vehicle velocity due to rotation. Acceleration terms are given in the following equations:

$$\text{Centripetal acceleration} = \omega_e \times (\omega_e \times p_{ECIC}) \tag{9.10}$$

$$\text{Tangential acceleration} = \omega_e \times v_b \tag{9.11}$$

Rotation also introduces an initial velocity relative to the earth's center for the vehicle before launch. Since the earth's surface rotates with respect to ECIC coordinates, a missile at rest before launch still has velocity with respect to the ECIC frame. The equation used to calculate this initial missile velocity is given in the following:

$$v_{init} = \omega_e \times p_{ECIC} \tag{9.12}$$

9.1.1.3 Oblateness Effects, Nonuniform Gravity

The latitude angles extending from the earth's center are different for an oblate earth model compared to a spherical model [4], and thus, various equations involving those angles must be defined. As shown in Figure 9.5 [3], there are two latitudes for an oblate model, the geocentric latitude from the earth's center, identified as μ_c, and the geodetic latitude, μ_d, on the semimajor axis directly below the earth's surface at the specified point. For a spherical model, these two were equivalent since a direction perpendicular to the earth's surface at any given point always intersected the earth's center. The term *latitude* on an oblate body, such as the earth, refers to the geodetic value.

The coordinate pair (v_s, z_s) refers to the earth surface geodetic subvehicle point. The coordinate pair (v_p, z_p) refers to the vehicle point at a geodetic altitude, H, above the surface. The oblate earth model also introduces a flattening factor, f, which describes the deviation of the geoid from a perfect sphere or the ellipticity of the earth and can significantly influence calculations and equations having to do with the flight equations of motion. One effect of an

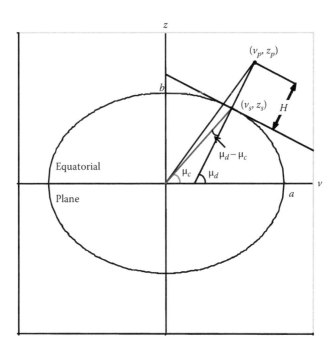

FIGURE 9.5
Illustration of geodetic versus geocentric latitude.

oblate earth is a nonuniform gravitational field [4]. In a spherical model, the gravity field vector is a constant toward the center of the earth regardless of latitude and longitude. With an oblate model, gravity is highly dependent on latitude yet still independent of longitude, assuming the model is flattened at the poles. Kaplan [4] gives the earth gravitational potential function $\vec{V}$ as 9.8:

$$\vec{V} = \frac{GM}{r}\left[1 + \sum_{n=2}^{n_{max}}\sum_{m=0}^{n}\left(\frac{a}{r}\right)^n \bar{P}_{n,m}(\sin\mu_c)\left(\bar{C}_{n,m}\cos m\ell + \bar{S}_{n,m}\sin m\ell\right)\right] \quad (9.13)$$

The term GM is the universal gravitational constant, r represents the distance from the earth center, and the subscripts (m, n) represent the order and degree, respectively, of the associated Legendre function (P) and the gravitational coefficients (C, S). The overbar represents normalization according to the following:

$$\left[\frac{(n+m)!}{(n-m)!\,(2n+1)^k}\right]^{\frac{1}{2}} \cdot (\bullet) \quad (9.14)$$

The associated Legendre function is defined in the following equation:

$$P_{n,m}(\sin\mu_c) = (\cos\mu_c)^m \cdot \frac{d^m}{d(\sin\mu_c)^m} \cdot [P_n(\sin\mu_c)] \qquad (9.15)$$

The Legendre polynomial is defined in the following:

$$P_n(\sin\mu_c) = \frac{1}{2^n n!} \cdot \frac{d^n}{d(\sin\mu_c)^n} (\sin^2\mu_c - 1)^n \qquad (9.16)$$

A perfectly oblate earth model that is symmetric about the z-axis has no zonal harmonics beyond a degree, $n = 2$. The larger the value, n, the less impact the magnitude of each coefficient has on the overall equation for gravity potential. For sensible atmospheric flight and ballistic flights less than 2000 km in range, a value of $n = 2$ should be reasonable. The so-called, J_2, Jeffry constant typically implies that an oblate spheroid earth model is being assumed. With the J_2 assumption, Equation 9.13 can be simplified and is rewritten in the following:

$$\vec{V} = \frac{GM}{r}\left[1 + \left(\frac{a}{r}\right)^2 \bar{P}_{2,0}(\sin\mu_c)(\bar{C}_{2,0})\right] \qquad (9.17)$$

The $(-\bar{C}_{2,0})$ term in this equation is in normalized form and can be related to the J_2 term and is defined [5] as follows for an order value $m = 0$, when the constant $\bar{C}_{n,m}$ is assumed as shown in the following equation:

$$J_n = -\bar{C}_{n,0}\sqrt{2 \cdot n + 1} \qquad (9.18)$$

Therefore, we can calculate the J term with $n = 2$ and $m = 0$ in the following equation:

$$J_2 = 0.0010826269 \qquad (9.19)$$

Using the definition for $\bar{P}_{2,0}(\sin\mu_c)$, Equation 9.17 reduces to the following equation, with $\mu_c = \tan^{-1}\left[(1-f)^2 \tan(\mu_d)\right]$ [1]:

$$\vec{V} = \frac{GM}{r}\left[1 - 0.5 \cdot J_2\left(\frac{a}{r}\right)^2 \cdot \{3\sin^2(\mu_c) - 1\}\right] \qquad (9.20)$$

The gradients of the potential function are evaluated in ECIC coordinates to give the following x, y, and z components of gravity where XE, YE, and ZE are the vehicles' respective ECIC coordinates:

$$G_X = \frac{GM}{r^2}\left[1+\frac{3J_2}{2}\left(\frac{a}{r}\right)^2\left(1-5\sin^2\mu_c\right)\right]XE/r$$

$$G_Y = \frac{GM}{r^2}\left[1+\frac{3J_2}{2}\left(\frac{a}{r}\right)^2\left(1-5\sin^2\mu_c\right)\right]ZE/r \quad (9.21)$$

$$G_Y = \frac{GM}{r^2}\left[1+\frac{3J_2}{2}\left(\frac{a}{r}\right)^2\left(1-5\sin^2\mu_c\right)\right]ZE/r$$

9.1.1.4 Geodetic and Geocentric Latitude Relationship

The mathematical relationship of the geocentric and geodetic latitude is obtained. Refer to Figure 9.5. The two points, one on the earth's surface (v_s, z_s) and one directly above that point (v_p, z_p), can be described in two separate but related methods. The relation between the two geodetic and geocentric coordinates is the flattening factor that gives the relation between the two angles. In the following equations, we will derive the relationship between geodetic and geocentric latitude and produce another equation to calculate vehicle position in ECIC coordinates given its longitude and geodetic latitude and altitude from Sooy [3]. From Figure 9.5, the following equation is defined:

$$\tan(\mu_d) = \frac{z_s}{v_s} \quad (9.22)$$

The equation of an ellipse from Figure 9.5 is provided in the following equation:

$$\frac{v^2}{a^2} + \frac{z^2}{b^2} = 1 \quad (9.23)$$

The earth equatorial radius is a, and b is the polar radius. If we set $b = (1-f)a$, we can rearrange Equation 9.23 and obtain the following equation:

$$z^2 = (1-f)^2(a^2 - v^2) \quad (9.24)$$

The slope of the line tangential to the ellipse for any given v and z is given in the following:

$$m = \frac{dz}{dv} = -\frac{v}{z}(1-f)^2 \quad (9.25)$$

The slope of the line perpendicular to Equation 9.25 is the negative reciprocal. The perpendicular slope passing through (v_s, z_s) is then given by the following equation:

$$m_\perp = \frac{z_s}{v_s(1-f)^2} \tag{9.26}$$

The equation of the slope of Equation 9.26 is also the equation for $\tan(\mu_c)$. Thus, with substitution, we have a relation between $\tan(\mu_c)$ and $\tan(\mu_d)$ given as follows:

$$\tan(\mu_d) = m_\perp = \frac{\tan(\mu_c)}{(1-f)^2} = \frac{z_s}{v_s(1-f)^2} \tag{9.27}$$

9.1.1.5 Geodetic to Geocentric Latitude: Vehicle Position

Most often, vehicle location is provided in longitude, geodetic latitude, and altitude. Flight performance computation is most often accomplished in geocentric or ECIC coordinates. To calculate vehicle position in ECIC coordinates given geodetic coordinates, we will solve Equation 9.27 for z_s and substitute this into the equation for the ellipse to give us the following equations:

$$\frac{v_s^2}{a^2} + \frac{v_s^2(1-f)^4 \tan^2(\mu_d)}{(1-f)^2 a^2} = 1 \tag{9.28}$$

Solving for v_s (magnitude) gives

$$v_s = \frac{a}{\sqrt{1+(1-f)^2 \tan^2(\mu_d)}} \tag{9.29}$$

Rearranging v_s (sign and magnitude)

$$v_s = \frac{a\cos(\mu_d)}{\sqrt{\cos^2(\mu_d)+(1-f)^2 \sin^2(\mu_d)}} \tag{9.30}$$

Equation 9.30 can be simplified to the form given in Equation 9.31. Substituting v_s and f into Equation 9.27 and solving z_s provide the subvehicle point on the oblate earth:

$$v_s = \frac{a\cos(\mu_d)}{\sqrt{1+f(f-2)\sin^2(\mu_d)}} \tag{9.31}$$

Assuming the vehicle is at some altitude (H) above the earth surface, the geodetic subvehicle point (v_s, z_s) will need to be adjusted to locate the vehicle in ECIC coordinates. From [4], the altitude-adjusted differential values for (v_s, z_s) are defined in the following equations:

$$\Delta v = H \cos(\mu_d) \tag{9.32}$$

$$\Delta z = H \sin(\mu_d) \tag{9.33}$$

The coordinates of the elevated point (v_p, z_p) are from [4] and given in the following equations:

$$v_p = v_s + \Delta v = \frac{a \cos(\mu_d)}{\sqrt{1 + f(f-2)\sin^2(\mu_d)}} + H \cos(\mu_d) \tag{9.34}$$

$$z_p = z_s + \Delta z = v_s(1-f)^2 \tan(\mu_d) + H \sin(\mu_d) \tag{9.35}$$

The ECIC coordinates of a vehicle can then be found by using the following equations when given only longitude (ℓ), geodetic latitude (μ_d), and geodetic altitude (H):

$$x_{ECI} = v_p \cos(\ell) \tag{9.36}$$

$$y_{ECI} = v_p \sin(\ell) \tag{9.37}$$

$$z_{ECI} = z_p = v_s(1-f)^2 \tan(\mu) + H \sin(\mu) \tag{9.38}$$

To transform the launch point (launch frame coordinates) and multiple vehicle positions to ECIC coordinates, an additional transformation will be required due to the oblate earth representation. This new transformation results from the difference in the angle between geodetic and geocentric latitude. For the spherical earth case, only three rotations were involved. Notice in Figure 9.5 that the oblate earth model now requires a fourth rotational transformation to align the vehicle's downward z-axis vector with the earth's center. The angle between the line perpendicular to the surface at a given point and the line from that point to the center of the earth is equal to the difference between the geodetic and geocentric latitude angles. This additional rotation is accomplished with another transformation matrix and accomplished in the same manner as previous transformation matrices.

9.1.1.6 Latitude, Longitude, and Altitude Calculation

To compute latitude, longitude, and altitude during flight is often required when designing midcourse guidance or when simply examining time–space correlated flight performance. Assume the engagement computational tool begins with latitude, longitude, and altitude of the launch frame with respect to the ECIC system. The initial target and interceptor position are acted on computationally in geocentric launch coordinates, but it is desired to reference those time–space correlated interceptor positions to geodetic assets. Thus, it is necessary to convert those changes in launch coordinates to ECIC positions using transformations and then calculate the new geodetic latitude, longitude, and altitude of each asset. These new values can then be used by the guidance algorithm(s) to direct the interceptor to the target and enable the user to correlate the trajectory of each vehicle in geodetic space. From [4], the following equations are given, where XE, YE, and ZE are the missile position in ECIC coordinates, e^2 is the eccentricity squared, and R_Θ is the equatorial radius of the earth:

$$\mu_d = \arctan\left[\frac{ZE}{(1-e^2)\sqrt{XE^2 + YE^2}}\right] \quad (9.39)$$

$$\ell = \arctan\left[\frac{YE}{XE}\right] \quad (9.40)$$

$$H = \sqrt{\frac{XE^2 + YE^2 + ZE^2}{1 + e^2 \sin^2(\mu)(e^2 - 2)}} - \frac{R_\Theta}{\sqrt{1 - e^2 \sin^2(\mu)}} \quad (9.41)$$

The assumed values for the WGS-84 ORE model [5] are defined as the following parameters and values. The first four parameters are the fundamental ones that can be used to recalculate all other coefficients as updates are required:

where
 R_Θ is the equatorial radius of the earth = 6.378137 * 10^6 m
 ω_Θ is the earth's angular rotational rate = 7.29211585530 * 10^{-5} rad/s
 $\bar{C}_{2,0}$ = −484.1654663 * 10^{-6}
 μ = 3.896004418 * 10^5 km³/s²
 e^2 is the eccentricity squared = 0.006694385000
 f is the flattening factor = 0.003352813000

The ORE provides a realistic ellipsoidal model to more accurately evaluate the flight performance of the AMD system and interceptor and specifically may be required when developing midcourse guidance designs. The earth's rotation introduces several new calculations including centripetal and tangential acceleration and time-dependent transformation matrices. Oblateness introduces a nonuniform gravity field and various new equations to calculate latitude, longitude, and altitude. A comprehensive treatment of earth models, coordinate systems, and transformations can be found in Valado [5].

9.1.1.7 Forces and Moments and Equations of Motion

We define aerodynamic forces and moments consistent with AIAA, R-004-1992 [6]. The aerodynamic reference system is defined in Figure 9.6. Normal, side, and axial forces (N_F, Y_F, A_F) and rolling, pitching, and yawing moments (L_m, M_m, N_m) are functions of independent variables angle of attack (α), angle of sideslip (β), aerodynamic roll angle (ϕ), Mach number and control surface deflection angles (δ), and the control surface aerodynamic incidence angle in the A-plane (i') and B-plane (i). Subsequently, it is necessary to define the relationships between the angles and the body axis velocity components to ensure proper interpretation of aerodynamic requirements.

The mathematic relationships in the following equation hold for Figure 9.6:

$$\alpha = \tan^{-1}(w/u); \quad \beta = \tan^{-1}(v/u); \quad \phi = \tan^{-1}(w/u)$$

$$\tan\alpha_T = \sqrt{\frac{(v^2+w^2)}{u^2}} = \sqrt{\tan^2\alpha + \tan^2\beta} \qquad (9.42)$$

$$V = \sqrt{u^2 + v^2 + w^2}$$

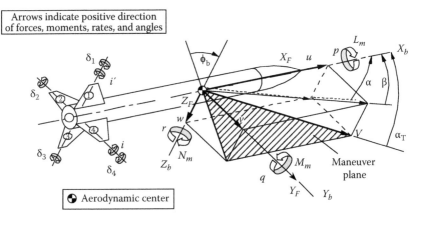

FIGURE 9.6
Vehicle aerodynamic reference system.

Physics and Mathematics of AMD Design and Analysis

The total velocity vector is defined as V, and the total angle of attack is defined as α_T.

Two reference frames are of interest before developing the complete six-degree-of-freedom (6DOF) equations of motion (EOM). We should assume that there is a locally level vehicle (LLV) carried reference frame that is oriented to the earth surface fixed Cartesian (ESFC) reference system, and we should assume that there is a missile body fixed reference system (B). Euler angles (ψ_M, Θ_M, Φ_M) represent the yaw, pitch, and roll orientation angles of B relative to LLV.

The reader should refer to Etkin [2, pp. 114–117] for details concerning this section. Vehicle body relative Euler angles (yaw, ψ_b; pitch, Θ_b; roll, ϕ_b) are normally employed to measure the angular displacement of the vehicle body attitude relative to the ESFC or LLV frame and will be followed here. Once the body Euler angles are found, they are used in a direction cosine matrix to compute the remaining required flight relationships.

The rate of change of the Euler angles between the ESFC/LLV frame axes and the vehicle body axes is defined by the Euler rate equations in terms of the body rotational rates (p_b, roll rate; q_b, pitch rate; r_b, yaw rate). The body rates are typically measured by the inertial reference unit (IRU) with rate gyroscopes. Homing interceptors are typically roll attitude stabilized with the exception of some shorter-range point defense systems. Moreover, as long as a relatively *tight* roll attitude control is maintained, the roll rate can be assumed zero for roll attitude–stabilized systems and the Euler rate equations can be simplified by setting $\dot{\Phi}_b = 0$.

The LLV to vehicle body axis direction cosine matrix $[A]$ is defined in terms of the Euler angles and given in Equation 9.43:

$$[A]_{LLV}^{B} = \begin{bmatrix} a_{11} & a_{12} & a_{13} \\ a_{21} & a_{22} & a_{23} \\ a_{31} & a_{32} & a_{33} \end{bmatrix} \quad (9.43)$$

$a_{11} = \cos\Theta_b \cos\Psi_b$

$a_{12} = \cos\Theta_b \sin\Psi_b$

$a_{13} = -\sin\Theta_b$

$a_{21} = \sin\Phi_b \sin\Theta_b \cos\Psi_b - \cos\Phi_b \sin\Psi_b$

$a_{22} = \sin\Phi_b \sin\Theta_b \sin\Psi_b + \cos\Phi_b \cos\Psi_b \quad (9.44)$

$a_{23} = \sin\Phi_b \cos\Theta_b$

$a_{31} = \cos\Phi_b \sin\Theta_b \cos\Psi_b + \sin\Phi_b \sin\Psi_b$

$a_{32} = \cos\Phi_b \sin\Theta_b \sin\Psi_b - \sin\Phi_b \cos\Psi_b$

$a_{33} = \cos\Phi_b \cos\Theta_b$

The $[A]_{LLV}^B$ matrix is an orthogonal transformation, and therefore, the direction cosine matrix to reverse the direction and go from B to LLV frame is its transpose and given in the following equation:

$$[A]_B^{LLV} = \begin{bmatrix} a_{11} & a_{21} & a_{31} \\ a_{12} & a_{22} & a_{32} \\ a_{13} & a_{23} & a_{33} \end{bmatrix} \quad (9.45)$$

The LLV velocity is found by multiplying the $[A]_B^{LLV}$ matrix and the vehicle body frame velocity (u, v, w) components shown in the following equations:

$$\begin{bmatrix} \dot{R}_{bx} \\ \dot{R}_{by} \\ \dot{R}_{bz} \end{bmatrix} = [A]_B^{LLV} \begin{bmatrix} u \\ v \\ w \end{bmatrix} \quad (9.46)$$

$$\dot{R}_{bx} = ua_{11} + va_{21} + wa_{31}$$
$$\dot{R}_{by} = ua_{12} + va_{22} + wa_{32} \quad (9.47)$$
$$\dot{R}_{bx} = ua_{13} + va_{23} + wa_{33}$$

It is the purpose of flight analysis to solve the earlier set of equations to understand the flight behavior of the interceptor under boost, midcourse guidance, or terminal homing to assess design alternatives. This is typically formulated in an engagement simulation where a set of parallel target equations exist, and they are solved relative to one another. It is instructive here to first establish the mechanization to solve the primary vehicle set of equations, and then generality applies to characterize a target in a similar fashion if this detail is warranted.

We first establish the effect that the rotating earth has on the vehicle body. Assuming an ORE model, we establish the following equation as the set of rotational rates imparted on the vehicle body axes from the earth:

$$p_b^E = \omega_e \cdot (a_{11} \cdot \cos \mu_d - a_{13} \cdot \sin \mu_d)$$
$$q_b^E = \omega_e \cdot (a_{21} \cdot \cos \mu_d - a_{23} \cdot \sin \mu_d) \quad (9.48)$$
$$r_b^E = \omega_e \cdot (a_{31} \cdot \cos \mu_d - a_{33} \cdot \sin \mu_d)$$

The resulting translational force equations can now be developed along the body axes. They are given in the following equation. The applied forces along the body axes include aerodynamic, propulsion, and environmental (wind) effects. The ORE gravity vector components $g_{X,\,Y,\,Z}$ are along the LLV frame and rotated using the body Euler angles to align with the vehicle body:

$$\dot{u} = \frac{X_F}{m} - g_X \cdot \sin\theta_b + \left(r_b^E + r_b\right)\cdot v - \left(q_b^E + q_b\right)\cdot w$$

$$\dot{v} = \frac{Y_F}{m} - g_Y \cdot \cos\theta_b \cdot \sin\phi_b + \left(p_b^E + p_b\right)\cdot w - \left(r_b^E + r_b\right)\cdot u \quad (9.49)$$

$$\dot{w} = \frac{Z_F}{m} - g_Z \cdot \cos\theta_b \cdot \cos\phi_b + \left(q_b^E + q_b\right)\cdot u - \left(p_b^E + p_b\right)\cdot v$$

Rotational equations of motion are developed assuming rigid-body dynamics and that a set of coordinate axes where the principal axes exist and the inertia matrix is a diagonal. Therefore, there are no products of inertia and the inverse of the inertia matrix is also a diagonal and Euler's equations of motion hold. Moments or torques are the result of force vectors applied about (at a distance from) the center of gravity and the body responds to in the form of angular accelerations. These vehicle body rotational equations of motion hold regardless of the earth model implemented and are given in the following equation:

$$\dot{p}_b = \frac{L_m}{I_{XX}} + \frac{(I_{YY} - I_{ZZ})\cdot q_b r_b}{I_{XX}}$$

$$\dot{q}_b = \frac{M_m}{I_{YY}} + \frac{(I_{ZZ} - I_{XX})\cdot r_b p_b}{I_{YY}} \quad (9.50)$$

$$\dot{r}_b = \frac{N_m}{I_{ZZ}} + \frac{(I_{XX} - I_{YY})\cdot p_b q_b}{I_{ZZ}}$$

The Euler angle rotation sequence is not commutative, and it is important to retain whatever sequence throughout the flight dynamics derivation and analysis. The rotation sequence is given here as (yaw, pitch, roll):

$$\begin{bmatrix} P \\ Q \\ R \end{bmatrix} = \begin{bmatrix} p_b \\ q_b \\ r_b \end{bmatrix} - [A]_B^{LLV} \begin{bmatrix} (\omega_e + \dot{\ell})\cos\mu_d \\ -\dot{\mu}_d \\ -(\omega_e + \dot{\ell})\sin\mu_d \end{bmatrix} \quad (9.51)$$

$$\dot{\Psi}_b = \frac{(Q\sin\Phi_b + R\cos\Phi_b)}{\cos\Theta_b}$$

$$\dot{\Theta}_b = Q\cos\Phi_b - R\sin\Phi_b \quad (9.52)$$

$$\dot{\Phi}_b = P + (Q\sin\Phi_b + R\cos\Phi_b)\tan\Theta_b$$

The Euler angle approach used to define vehicle orientation in space contains a singularity by inspection of the previous equation. As the second, or pitch rotation (Θ), approaches ±90°, the yaw (Ψ) and roll (Φ) Euler angle rates become undefined. To eliminate this problem, flight dynamics has evolved to using quaternions [1] to compute body attitude.

Quaternions are mathematically defined but have no physical meaning and in themselves have no usefulness in the study of flight dynamics. Quaternions are therefore only useful from a computational perspective. They must be converted back to Euler angles before any subsequent analysis can be accomplished. Specifically, quaternions are mathematical functions of the direction cosines of the vehicle orientation.

The initial vehicle orientation needs to be specified in Euler angles and then are converted to four quaternion values (e_0, e_1, e_2, e_3) that are used to initialize the quaternion rates when combined with the body rates. These quaternion rates are then integrated and converted back to Euler angles for interpretation and to begin the next computational iteration. The sequence for implementing quaternions is given in the following equations:

$$e_0 = \frac{1}{2}[\cos(\psi_b)\cdot\cos(\theta_b)\cdot\cos(\phi_b) + \sin(\psi_b)\cdot\sin(\theta_b)\cdot\sin(\phi_b)]$$

$$e_1 = \frac{1}{2}[\cos(\psi_b)\cdot\cos(\theta_b)\cdot\cos(\phi_b) - \sin(\psi_b)\cdot\sin(\theta_b)\cdot\sin(\phi_b)] \quad (9.53)$$

$$e_2 = \frac{1}{2}[\cos(\psi_b)\cdot\sin(\theta_b)\cdot\cos(\phi_b) + \sin(\psi_b)\cdot\cos(\theta_b)\cdot\sin(\phi_b)]$$

$$e_3 = \frac{1}{2}[-\cos(\psi_b)\cdot\sin(\theta_b)\cdot\sin(\phi_b) + \sin(\psi_b)\cdot\cos(\theta_b)\cdot\cos(\phi_b)]$$

$$\dot{e}_0 = \frac{1}{2}[e_1 P + e_2 Q + e_3 R]$$

$$\dot{e}_1 = \frac{1}{2}[e_0 P + e_2 R + e_3 Q]$$

$$\dot{e}_2 = \frac{1}{2}[e_0 Q + e_3 P - e_1 R] \quad (9.54)$$

$$\dot{e}_3 = \frac{1}{2}[e_0 R + e_1 Q + e_2 P]$$

$$\psi_b = \tan^{-1}\left[\frac{2\cdot(e_1 e_2 - e_0 e_3)}{e_0^2 + e_1^2 - e_2^2 - e_3^2}\right]$$

$$\theta_b = \sin^{-1}[-2\cdot(e_1 e_3 - e_0 e_2)] \quad (9.55)$$

$$\varphi_b = \tan^{-1}\left[\frac{2\cdot(e_2 e_3 - e_0 e_1)}{e_0^2 + e_3^2 - e_1^2 - e_2^2}\right]$$

9.2 Target and Clutter Returns

Clutter is a source of noise that interferes with target detection. By definition, clutter is any kind of unwanted echoes that originates from raindrops, birds, water, and ground surfaces. The occurrence of clutter may be a serious problem since it could result in echoes stronger than the intended target. It is often possible to discard much of the clutter by simply ignoring echoes from slow-moving objects using a Doppler gate, that is, by removing echoes not having the expected frequency. Using an appropriate frequency for the radar will also help decrease clutter since the reflectivity is frequency dependent. It is not always realistic to remove all clutter, and possible clutter must be taken into account in some real-life situations. Fortunately though, clutter can be removed with good results when using Doppler techniques and slow or stationary surface returns can be neglected.

9.2.1 Radar Returns

The radar range equation is used to determine both the target and clutter powers returned from a given range. The monostatic power returned from a target or clutter is a function of the radar parameters, range, and radar cross section as follows:

$$P_r(r) = \frac{P_t G_t A_e \sigma}{[4\pi]^2 r^4 L} \quad (9.56)$$

where
  P_t is the peak transmitter power
  G_t is the transmit gain of the radar antenna
  A_e is the effective receiving area of the radar antenna
  σ is the radar cross section of the target or clutter
  r is the slant range of the target or clutter
  L is total transmit, receive, and processing losses

A complete and rigorous derivation of the monostatic radar range equation is presented by Blake [7]. For distributed clutter, the clutter radar cross section is also a function of range. The following sections describe how the clutter radar cross section can be estimated for both surface and volume clutter.

9.2.2 Surface Clutter Returns

The geometry for a simple surface clutter model is shown in Figure 9.7 for a sea/land environment where mountains rise from the seashore.

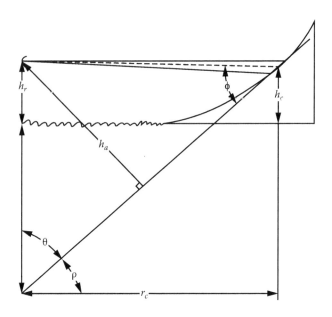

FIGURE 9.7
Surface clutter geometry for a sea–mountainous land interface.

The model is general in the sense that it accounts for rising terrain overland when determining the surface area intercepted by the main beam of the radar antenna at height h_r. The range of the surface clutter patch from the radar is r_c. The radar antenna and clutter heights with respect to a mean sea-level reference are h_r and h_c, respectively. The perpendicular distance from the radar to a line tangent to the clutter surface at the point where the antenna main beam axis intercepts the illuminated patch is the adjusted height, h_a, of the radar for an equivalent flat land scenario. In the model, the main beam axis is always assumed to intersect the illuminated clutter patch at each range.

Using the model geometry, the surface length of the clutter patch in both the azimuth and elevation extents of the antenna main beam can be determined. The clutter patch surface area is computed assuming the intersection of the main beam and clutter surface is an elliptical surface. The clutter surface length intercepted in the azimuth direction is a function of the antenna azimuth beamwidth, while the clutter surface length intercepted in the elevation direction is a function of either the pulse width or the antenna elevation beamwidth depending on the specific geometry. When the elevation surface length is a function of the pulse width, this is referred to as a *pulse-limited* clutter patch. *Beam-limited* clutter patch refers to the case of functional dependency on the antenna beamwidth. Beam limiting usually occurs for close-range clutter, with a transition to pulse limiting with increasing clutter range.

Physics and Mathematics of AMD Design and Analysis 227

To calculate the elevation surface length intercepted by the antenna beam for the pulse-limited case, the angle φ must be determined. The elevation surface length intercepted can be determined from

$$\zeta_{ep} = 1/2 c\tau \sec\varphi \qquad (9.57)$$

where
ζ_{ep} is the elevation direction surface length intercepted
c is the speed of light
τ is the pulse width

The angle φ is also necessary to compute the surface length intercepted for the beam-limited case, that is,

$$\zeta_{eb} = r_c \theta_E \csc\varphi \qquad (9.58)$$

where θ_E is the antenna 3 dB elevation beamwidth. When the pulse-limited length, ζ_{ep}, is greater than the beam-limited length, the surface clutter is beam limited. The clutter patch surface area is given for both the following cases:

$$S_p = \zeta_{ep} r_c \theta_E \quad \text{pulse limited} \qquad (9.59)$$

$$S_b = \pi/4 \zeta_{eb} r_c \theta_A \quad \text{beam limited} \qquad (9.60)$$

The surface clutter patch area is a truncated elliptical region for the pulse-limited case and elliptical for the beam-limited case.

The radar cross section of the distributed clutter illuminated can be estimated:

$$\sigma_c = S\sigma_o \qquad (9.61)$$

where σ_o is the mean value of the areal backscatter coefficient for the clutter. Typical values for the areal backscatter coefficient for different clutter types can be found in the literature [9].

Once the clutter patch radar cross section is determined, the received clutter power can be determined by substitution of the clutter patch radar cross section, σ_c, into the radar range equation. The clutter power returned for the pulse-limited case decays as the inverse cube of range. At the transition range, the beam-limited and pulse-limited clutter patch areas are equal, providing a smooth transition.

9.2.3 Volume Clutter Returns

For volume clutter such as rain, it is necessary to determine the illuminated clutter volume at each range. The illuminated clutter volume can be treated as two separate cases: (1) the beam (radar range–angle cell) is completely filled with clutter and (2) the beam is partially filled with clutter. When the beam is filled with clutter, the beam shape in angle and the pulse width in slant range bound the clutter volume. Assuming an elliptically shaped antenna beam, the clutter volume illuminated for the beam-filled case can be expressed as

$$V = \pi/4(r_c\theta_E)(r_c\theta_A)(c\tau/2) \tag{9.62}$$

Since the gain of the transmit and receive beams is not uniform over the areal beamwidth of the antenna, the aforementioned expression is usually reduced by the factor 2(ln 2) assuming a Gaussian two-way antenna pattern [9], that is,

$$V = \pi/(16(\ln 2))(r_c\theta_E)(r_c\theta_A)(c\tau) \tag{9.63}$$

The radar cross section of the volume clutter illuminated is estimated as

$$\sigma_c = V\sigma_v \tag{9.64}$$

where σ_v is the mean value of the volumetric backscatter coefficient. Typical values for the volumetric backscatter coefficient for various clutter types can be found in the literature [8]. Once the illuminated clutter volume radar cross section is determined, the received clutter power can be determined from the radar range equation. Beam-filled volume clutter returns decay as the inverse square of range.

When the antenna beam (range–angle cell) is not filled with clutter, the clutter volume is no longer bounded in elevation by the beam shape. For non-beam-filled clutter, two phenomena can occur simultaneously or individually: (1) the beam shape exceeds the clutter maximum height in elevation and/or (2) part of the beam intercepts the surface of the earth. Figures 9.8 and 9.9 illustrate the two cases in a cross-sectional view perpendicular to the main beam axis assuming an elliptically shaped two-way antenna areal beamwidth.

To determine the cross-sectional area illuminated, it is necessary to know the areas of the triangular region and elliptic sector shown in Figure 9.9. The illuminated area of the truncated semiellipse of Figure 9.9 can be determined straightforwardly when the height, h, from the main beam axis to the truncation point is known. The base length of the triangular sector, g, can be determined from geometry as

$$g = [a^2 - h^2(a/b)^2]^{1/2} \tag{9.65}$$

Physics and Mathematics of AMD Design and Analysis

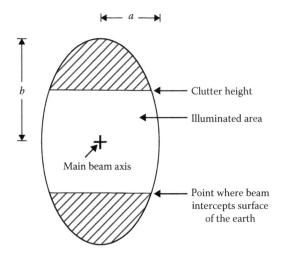

FIGURE 9.8
Illustration showing the cross-sectional view of non-beam-filled volume clutter.

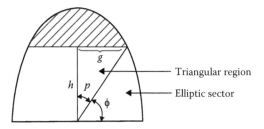

FIGURE 9.9
Illustration showing the semiellipse geometry used to determine the cross-sectional area of the illuminate clutter volume.

$$a = r_c \theta_A / 2 \tag{9.66}$$

$$b = r_c \theta_E / 2 \tag{9.67}$$

where a and b are the ellipse semiaxis lengths. At this point, the angle φ can be determined as

$$\varphi = \pi/2 - \tan^{-1}(g/h) \tag{9.68}$$

Next, the area of the elliptic sector is determined by computing the incremental area of the ellipse as a function of the angle φ. This elliptic sector is determined by integrating in cylindrical coordinates. The result of this integration results in the area of the elliptic sector as

$$A_E = (ab)/2 \tan^{-1}[(a/b)\tan \varphi] \tag{9.69}$$

The semi-ellipse area illuminated is just twice the sum of the areas of the triangular and elliptic sector regions:

$$A_{SE} = gh + (ab)\tan^{-1}[(a/b)\tan\varphi] \quad (9.70)$$

The illuminated volume of the partially filled beam can now be approximated as

$$V = \left(A_{SE}^{UPPER} + A_{SE}^{LOWER}\right)c\tau/(4\ln 2) \quad (9.71)$$

where the length of the truncated elliptic cylinder is $c\tau/2$. This result assumes that the antenna beam axis is parallel to the earth's surface at each clutter range.

9.2.3.1 Clutter Processing Considerations

Ambiguous clutter returns occur in an MTI or PD waveform when clutter exists beyond the unambiguous range of the waveform. The unambiguous range is the range associated with the duration of the pulse repetition interval (PRI) or time between successive pulses. For a PRI of 0.1 μs, the unambiguous range is 15 km. Ambiguous range clutter returns compete with targets in the unambiguous range interval for detection. All targets of interest are assumed to lie within the unambiguous range. In addition, the ambiguous or folded clutter return may not be present in each of the pulse returns processed. When this occurs, the clutter filter is no longer matched to reject the clutter return, resulting in decreased clutter rejection. This effect can be quite significant for clutter filters that only process several pulses.

The following sections discuss topics that directly relate to the ability of the clutter processing to reject ambiguous clutter effectively. The topics discussed include the clutter processing interval, maximum clutter range, fill pulses, and clutter rejection degradation.

9.2.4 Coherent Processing Interval

The coherent processing interval (CPI) can be best described by referring to a simple example. In this example, the scenario is limited to three range frames, which contain either a target or a clutter that is processed using three-pulse MTI. A range frame is simply a range interval equal to the unambiguous range, r_{ua}, of the waveform. The target and clutter scenario is shown in Figure 9.10 with the target denoted by the caret symbol and clutter denoted by the asterisks. For this example, point clutter is used; however, the results apply to distributed clutter as well. A target is located at a range of 0.5 times, the unambiguous range, and clutter is located at 1.5 and 2.5 times, the unambiguous range.

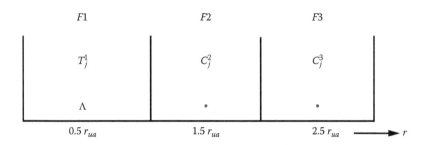

FIGURE 9.10
Example unambiguous clutter scenario.

The superscripts refer to the range frame where the clutter or target physically resides and subscripts refer to the pulse number within the burst from which the radar return is received. For a five-pulse MTI waveform (see Figure 9.10), the time coincident returns in the first three processing intervals are shown in Table 9.1, where the coherent processing interval covers one unambiguous range interval for three contiguous range frames. The number of the CPI corresponds to the pulse number where clutter processing begins, that is, CPI #1 begins at pulse 1, CPI #2 begins at pulse 2, and so on. The data are processed out to the unambiguous range of the last pulse in the burst, and typically, the first available coherent processing interval is used. It is observed that the mixture of target and clutter returns is vastly different depending on which CPI is selected for processing.

For example, in CPI #1, the target return is present in each pulse return; however, the clutter return from range frame #2 is in two returns and the clutter from range frame #3 is only in one return. Close examination of Table 9.1 reveals that the target appears in each range frame processed in CPI #1, that is, a target within the unambiguous range interval will always appear in each frame of CPI #1. Further, the number of missing clutter returns in the first available CPI is equal to the displacement in range frames of the CPI from the actual clutter range. Since missing clutter returns result in degraded clutter rejection, it is desirable to *fill* the coherent processing interval with clutter. The task of the clutter processing waveform designer is to provide an adequate number of fill pulses for the specified clutter scenario as illustrated

TABLE 9.1

Time Coincident Radar Returns for Three Pulses for Scenario

	CPI #1	CPI #2	CPI #3
Frame 1	T_1^1	$T_2^1 + C_1^2$	$T_3^1 + C_2^2 + C_1^3$
Frame 2	$T_2^1 + C_1^2$	$T_3^1 + C_2^2 + C_1^3$	
Frame 3	$T_3^1 + C_2^2 + C_1^3$		

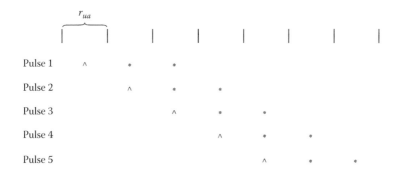

FIGURE 9.11
Radar returns from five coherent pulses for Table 9.1 scenario.

in Figure 9.11. Since real-world clutter environments can be quite complex, a set of waveforms may be required that can be adapted to the clutter environment to optimize clutter rejection and target detection.

9.2.5 Fill Pulses

Fill pulses are used to fill in the missing clutter returns. They are extra pulses in addition to the number required for a particular clutter filter. In this example case, three pulses are required with additional pulses used to fill in clutter returns. Assuming four pulses are transmitted and processing does not begin until after the second pulse is transmitted (CPI #2 processing), the time coincident returns are given in Table 9.2.

Note that CPI #2 (the first available CPI) is now filled except for a single missing clutter return in frame 1 at long range, C^3, which will cause some degradation in clutter rejection against the C^3 clutter component. If only the second time around clutter is present, then CPI #2 is the preferred processing interval; however, the third time around clutter can be effectively canceled by processing CPI #3 with a two-pulse MTI.

Now, consider a five-pulse burst where processing begins after pulse 3 is transmitted. This corresponds to the case of using two fill pulses. The time coincident radar returns are summarized in Table 9.3.

TABLE 9.2

Time Coincident Radar Returns for Four Pulses with One Fill Pulse Delay

	CPI #2	CPI #3	CPI #4
Frame 1	$T_2^1 + C_1^2$	$T_3^1 + C_2^2 + C_1^3$	$T_4^1 + C_3^2 + C_2^3$
Frame 2	$T_3^1 + C_2^2 + C_1^3$	$T_4^1 + C_3^2 + C_2^3$	
Frame 3	$T_4^1 + C_3^2 + C_2^3$		

Physics and Mathematics of AMD Design and Analysis

TABLE 9.3
Time Coincident Radar Returns for Five Pulses with Two Fill Pulses Delay

	CPI #3	CPI #4	CPI #5
Frame 1	$T_3^1 + C_2^2 + C_1^3$	$T_4^1 + C_3^2 + C_2^3$	$T_5^1 + C_4^2 + C_3^3$
Frame 2	$T_4^1 + C_3^2 + C_2^3$	$T_5^1 + C_4^2 + C_3^3$	
Frame 3	$T_5^1 + C_4^2 + C_3^3$		

Now, the clutter returns are completely filled in CPI #3. Clearly, for nth time around clutter, $n - 1$ fill pulses are needed to completely fill in the clutter in the first available processing interval after the fill pulses are transmitted.

9.2.6 Maximum Clutter Range

The maximum clutter range is defined as the maximum range at which a clutter return can be received and processed. This also corresponds to the minimum range at which a jammer can intercept and repeat a single pulse of the waveform that is received and processed by the radar. The maximum clutter range is proportional to the total number of pulses in the burst and is given as

$$r_{mc} = [c(M + FP)PRI]/2 = r_{ua}(M + FP) \tag{9.72}$$

where
FP is the number of fill pulses
M is the number of pulses coherently processed

This is the range beyond which clutter returns and responsive jamming are not received and processed by the radar.

9.2.7 Clutter Rejection Degradation

Range ambiguous clutter returns degrade clutter rejection when the ambiguous clutter returns are not completely filled. Specifically, one or more range frames may not contain the return(s) from a range ambiguous clutter source. When this occurs, one or more of the pulses do not contain the same clutter returns as the other pulses, resulting in reduced clutter rejection (the clutter notch depth is reduced). For example, in Figure 9.12, the frequency response of a five-pulse MTI filter with Chebyshev weighting is shown for 0 missing returns (1st time clutter), 1 missing return (2nd time clutter), and 2 missing returns (3rd time clutter).

Multiple time around clutter reduces the ability of the MTI filter to reject the low-velocity radar returns associated with the clutter that is slow moving or stationary. The low-velocity region of the clutter filter is sometimes referred to as the clutter notch. In general, the clutter may be distributed

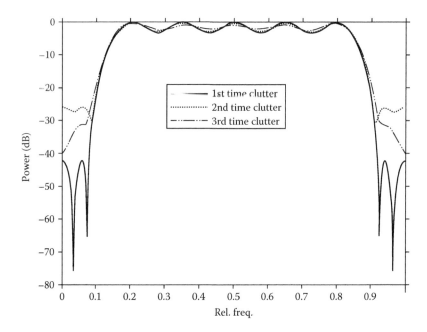

FIGURE 9.12
Effect of range ambiguous clutter on MTI clutter rejection capability.

in velocity and overlap the transition and/or passband of the clutter filter. When this is the case, the clutter cannot be completely rejected unless an MTI filter with a wider clutter notch is used.

9.2.8 Range–Velocity Visibility

A target is said to be visible in a given range–velocity cell if the signal-to-clutter pulse noise ratio at the detector meets or exceeds the required level to achieve a specified probability of detection. Range eclipsing and blind velocities result from implementation practices and ultimately determine the range–velocity visibility of a waveform, while ineffective clutter rejection can result in greatly degraded subclutter visibility.

Eclipsed ranges result from turning off or blanking the receiver while transmitting each pulse in the burst since it is difficult to obtain enough isolation to receive radar returns without degraded sensitivity during high-power transmission. Blind velocities occur because the clutter filter frequency response rejects targets that lie within the clutter notch. The effects of eclipsing and blind velocities can be reduced by the use of multiple bursts with different pulse repetition frequencies.

Loss in target visibility due to insufficient clutter rejection is very much dependent on the specific clutter environment. Percent target visibility is a figure of merit that is used in designing clutter rejection waveforms. Some range–velocity regions may be given a higher priority depending on the application and need to be weighted accordingly. The following paragraphs discuss eclipsing, blind velocities, and visibility in clutter in more detail.

9.2.9 Eclipsing

Eclipsing, in general, may result from at least two sources: (1) receiver blanking during high-power transmission and (2) when the radar antenna has nonreciprocal phase shifters such as in some phased array designs, the receiver is additionally blanked during the time it takes to switch the phase shifters from receive to transmit and then from transmit to receive. The total blanking time determines the eclipsed range interval:

$$r_e = c(\tau + t_{pc})/2 \tag{9.73}$$

where t_{pc} is the phase shifter cycling time. Eclipsing occurs each time a pulse in the burst is transmitted.

9.2.10 Blind Velocities

Blind velocities result because the clutter filter response repeats at multiples of the pulse repetition frequency (PRF). Since the unambiguous velocity is equal to the PRF, the blind velocity interval will repeat at multiples of the unambiguous velocity that is given as

$$v_{ua} = c/(2PRIf_c) \tag{9.74}$$

where f_c is the radar microwave frequency. The blind velocity interval is determined by the width of the notch region of the MTI filter.

9.2.11 Visibility in Clutter

The percent of the instrumented range–velocity space visible for a given waveform is a figure of merit often used to gauge performance. A target is visible if it is not located at an eclipsed range or blind velocity and the signal-to-clutter plus noise ratio is sufficient to achieve the required probability of detection. When the MTI waveform consists of more than one burst PRF in order to resolve velocity ambiguities and to effectively reduce or eliminate blind velocities, a particular target range–velocity combination is said to be visible if the requirements are met for one or more of the multiple PRFs. In other words, if the requirement is met for at least one PRF for a given target range–velocity combination, the target is visible.

Since some target range–velocity combinations may have a higher priority for detection than others, such as a high-velocity target at close range as opposed to a low-velocity target at long range, it may be desirable to weight the range–velocity space. For the general case of nonuniform weighting, the percent coverage can be written as

$$\%C = \frac{\Sigma W_v(r,v)}{\Sigma W(r,v)} \tag{9.75}$$

where W_v are the weights for the visible range–velocity space and the denominator is the sum of all weights over the entire space under consideration.

Various combinations of range and velocity weighting should provide a reasonable amount of flexibility. Three range weighting alternatives are considered. The range weightings are (1) linearly decreasing with range, (2) exponential decreasing, and (3) a combination of uniform and monotonically decreasing weights as shown in Figure 9.13.

The slope for the linearly decreasing range weighting is specifiable such that the range axis intercept point always occurs at the maximum range. For the case of exponential decay weighting, the rate of decay is specifiable. For combinational weighting, a uniform weight is applied out to specifiable range r_t and an exponentially decreasing weight is applied from range r_t to the maximum range.

The two velocity weightings considered are (1) uniform and (2) linearly increasing with velocity with the velocity intercept occurring at zero velocity. The range and velocity weights are assumed to be separable, so the combined weight is just their product:

$$W_{rv} = W_r W_v \tag{9.76}$$

This approach allows greater emphasis to be given to high-velocity, short-range targets that stress the capabilities of the radar and weapon system.

9.3 Seeker Noise Sources

9.3.1 Receiver Noise

Receiver noise is a range-dependent noise source. The effect of receiver noise and the guidance command generation needs to be considered and possibly filtered before seeker angle measurements can be used in a guidance law. Receiver noise comes from a number of different sources, both natural and man-made, and according to Nesline [10], receiver noise is the thermal noise produced at the radar receiver with range dependence governed by the radar range equation as shown in the following equation:

$$\sigma_{rn} = \sigma_{RN0} \cdot \frac{R_{TM}}{R_0} \tag{9.77}$$

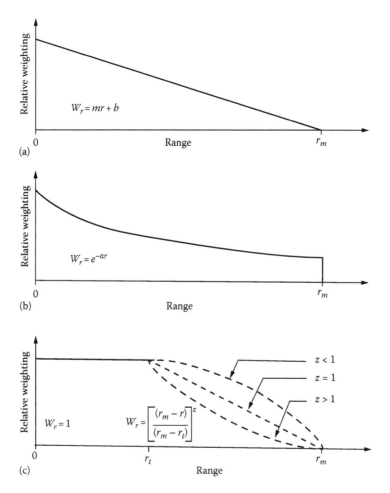

FIGURE 9.13
Three types of range weighting for % coverage determination: (a) linear weighting, (b) exponential weighting, and (c) uniform/exponential weighting.

where σ_{RN_0} is the receiver noise standard deviation at the reference range, R_0, given a receiver noise correlation time constant, T_{RN_0}. The white reference receiver noise angular error standard deviation is a function of its power spectral density (PSD) and is found using the following equation:

$$\sigma_{RN_0} = \sqrt{\frac{\Phi_{RN_0}}{2 \cdot T_{RN_0}}} \qquad (9.78)$$

R_{TM} is the time-dependent range between the target and the interceptor missile. Internal noise is primarily introduced in the early stages of the receiver, in that it passes through all amplifying stages of the receiver and will therefore dominate noise introduced later in the receiver. Other sources of radio signals are the ground, space, etc., but in airborne applications these are negligible compared to the internal noise [9].

9.3.1.1 Glint

Much has been written [10–15] on Glint and its associated effects on missile terminal homing accuracy and miss distance. Glint can be defined as the RCS fluctuation-induced apparent LOS angle and angle rate change. Glint is not dependent on the radar or specific tracking techniques (conical scan, monopulse, etc.) and is principally the result of coherent scattering from a complex target (e.g., body, wings, tail). When considering the reflections from a complex target, they are not simply from a point but are a collection of point scatterers all contributing to the reflections collected by the tracking radar.

Target motion in the form of maneuvering or stabilization motions (short or long period motions for example) contributes glint by changing the relative phase and magnitude relationship between each independent scatterer. The resultant effect of this motion is to cause an apparent random wander of the radar center of the target being tracked. The wander occurs in both angle and range and will have rate components. Glint is particularly a problem for seekers as it is a low-frequency phenomenon and usually falls in the passband of the servos driving the antenna and cannot be easily filtered. Compounding the problem is that as the intercept range closes to zero, the angle error increases infinitely per lateral fluctuation deviation resulting in the saturation of the servo system.

Glint, for complex targets with many scatters, is statistically characterized by James [16], Barton [17], and Garnell [18] as a Gaussian probability density function (PDF) with the mean and standard deviation of the spectrum tied to the target dimensions. According to Garnell, glint is largely a phenomenon in the target yaw plane and the RMS value can be characterized as 1/5 its wingspan, while James claims values between 1/3 and 1/6 are appropriate. Gordon [15] characterizes glint standard deviation as follows:

$$\sigma_{glint} = \frac{k \cdot L}{R} \tag{9.79}$$

where
k (0.1–0.2) is the proportionality constant from empirical study
L is the target wingspan
R is the range to target

Although Gordon predicts that glint follows a time-correlated, student-t distribution, he concludes that using a Gaussian distribution in an extended

Physics and Mathematics of AMD Design and Analysis

nonlinear Kalman filter is appropriate. Gordon should be consulted for a sophisticated approach to modeling glint.

Garnell proposes to model the glint spectrum as white noise passed through a first-order lag with correlation time constant (T_g) and shown in the following:

$$\Phi(\omega) = \frac{K_g^2}{1+\omega^2 T_g^2} \; m^2/rad/s \tag{9.80}$$

where
$K_g^2 = \pi \cdot k_g^2 / 2 \cdot T_g$ is the mean square value of the glint
T_g is typically in the numeric ranges between 0.1 and 0.25

Gordon, referencing Barton, expresses T_g as follows:

$$T_g \cong \frac{\lambda}{3.4 \cdot \omega_a \cdot L} \tag{9.81}$$

where
λ is the seeker radar RF wavelength
ω_a is the rate of change of the interceptor to target aspect angle (rad/s)
L is the span of the target across the line of sight

9.3.1.2 Radome Boresight Error

According to Nesline and Zarchan [13], the radome boresight error slope (RBS) is the rate of change of the refraction angle with the gimbal angle and can be expressed mathematically as

$$R = dr/d\theta_H \tag{9.82}$$

The radome slope R can be modeled according to the relationship developed from the empirical data and provided by Nesline and Zarchan [13] and shown here in Equations 9.83 and 9.84:

$$R = \pm R_T / 2 \tag{9.83}$$

$$R_T = \frac{2.83 \cdot (FR - 0.5) \cdot (1 + (2.35 \cdot \varepsilon \cdot B)^2)}{k_m \cdot ds \cdot \sqrt{\varepsilon - 1}} \tag{9.84}$$

The following radome material parameters and definitions [13] in Table 9.4 apply.

TABLE 9.4

Radome Material and Characteristic Parameters

Radome Dielectric Material Constant (Room Temperature)	Dielectric Value
Slip cast fused silica	3.5
RayceramIII™	4.8
Pyroceram 9606™	5.5
BeO	6.7
97% Al_2O_3	9.0
Variable	Definition
FR	Radome fineness ratio (radome length/radome diameter)
d_s	Antenna dish diameter
ε	Radome material dielectric constant
B	Deviation from design frequency
λ	Signal wavelength
k_m	Figure of merit
R_T	Maximum percent of swings of 90% of slopes

References

1. Stevens, B.L. and Lewis, F.L., *Aircraft Control and Simulation*, 2nd edn., John Wiley & Sons, Hoboken, NJ, 2003.
2. Etkin, B., *Dynamics of Atmospheric Flight*, John Wiley & Sons, New York, 1972.
3. Sooy, T.J., Description of Interceptor Flyout Model (IFM) upgrade to oblate, rotating earth model, TSC-W255-006/tjs, June 16, 2004.
4. Kaplan, M., *Modern Spacecraft Dynamics & Control*, John Wiley & Sons, New York, 1976, pp. 273–284.
5. Valado, D.A., *Fundamentals of Astrodynamics and Applications*, 2nd edn., Microcosm Press/Kluwer Academic Publishers, Dordrecht, the Netherlands, 2004.
6. AIAA, R-004-1992, *Recommended Practice for Atmospheric and Space Flight Vehicle Coordinate Systems*, Washington, DC, February 1992.
7. Blake, L., *Radar Range-Performance Analysis*, Artech House, Norwood, MA, 1986.
8. Nathanson, F.E., *Radar Design Principles*, McGraw-Hill, New York, 1969.
9. Skolnik, M., *Introduction to Radar Systems*, McGraw-Hill, New York, 1982.
10. Nesline, F.W. and Zarchan, P., Miss distance dynamics in homing missiles, in *AIAA Guidance and Control Conference Proceedings*, Paper No. 84-1844, Snowmass, CO, pp. 84–98, 1984.
11. Sandhu, G.S., A real time statistical radar target model, *IEEE Transactions on Aerospace and Electronic Systems*, AES-21(4), 490–507, July 1985.
12. Huynen, J.R., McNolty, F., and Hansen, E., Component distributions for fluctuating targets, *IEEE Transactions on Aerospace and Electronic Systems*, AES-11(6), 1316–1331, November 1975.

13. Nesline, W. and Zarchan, P., Radome induced miss distance in aerodynamically controlled homing missiles, in *AIAA Guidance Navigation and Control Conference Proceedings*, Paper No. 84-1845, pp. 99–115.
14. Borden, B., What is the radar tracking "Glint" problem and can it be solved, NAWCWPNS TP 8125, Naval Air Warfare Center Weapons Division, China Lake, CA, May 1993.
15. Gordon, N. and Whitby, A., Bayesian approach to guidance in the presence of glint, *AIAA Journal of Guidance, Control and Dynamics*, 22(3), 478–485, May–June 1999.
16. James, D.A., *Radar Homing Guidance for Tactical Missiles*, Macmillan Education, New York, 1986.
17. Barton, D.K., *Radar System Analysis*, Prentice Hall, Englewood Cliffs, NJ, 1964.
18. Garnell, P. and East, D.J., *Guided Weapon Control Systems*, Pergamum Press, Oxford, UK, 1977.

Acronyms

AAD	Area air defense
ADI	Air defense interceptor
ADM	Air defense missile
ADP	Architecture design proposal
AMD	Air and missile defense
AOA	Angle of attack
BAT	Battlespace assessment tool
BCW	Body–canard–wing
BEAT	Battlespace engineering assessment tool
BECO	Booster engine cut-off
BMD	Ballistic missile defense
BMP	Battle management processor
BMS	Battle management system
BT	Body–tail
BWT	Body–wing–tail
CCM	Configuration control management
CDR	Critical design review
CDS	Central defense system
CFA	Crossed-field amplifier
CLS	Communication link system
CONOPS	Concept of operations
CPI	Coherent processing interval
CW	Continuous wave
dB	Decibel
dBsm	dB square meter
DPO	Defense penetration options
DPT	Defense penetration technique
DRFM	Digital RF memory
ECC	Engagement control computer
ECEF	Earth-centered earth-fixed
ECIC	Earth-centered inertial Cartesian
ECM	Electronic counter measures
ECS	Engagement computer system
EM	Electromagnetic
EO	Electro-optical
ESFC	Earth surface fixed Cartesian
EW	Electronic warfare
FCS	Fire-control system
FOS	Family of systems
FOV	Field of view

FSS	Frequency selective surface
GHz	Gigahertz
GIA	Guidance integrated airframe
GIC	Guidance integrated control
GIN	Guidance integrated noise
GIT	Guidance integrated target
GLONASS	Global navigation satellite system
GPS	Global positioning system
GNC	Guidance, navigation, and control
HE	Heading error
HEW	High-energy weapon
IBW	Instantaneous bandwidth
ID	Identification
IFF	Identification friend or foe
IPT	Integrated product team
IR	Infrared
IRU	Inertial reference unit
ISR	Intelligence, surveillance, and reconnaissance
kW	Kilowatt
LAR	Lifecycle assessment reviews
LCIC	Launch-centered inertial cartesian
LLV	Locally level vehicle
LNA	Low noise amplifier
LOAL	Lock-on after launch
LOBL	Lock-on before launch
LOS	Line of sight
LR	Long range
m	Meter
MHz	Megahertz
µs	Microsecond
MITL	Man in the loop
MNS	Mission needs statement
MOE	Measure of effectiveness
MR	Medium range
MRBM	Medium-range ballistic missile
MTI	Moving target indicator
MW	Megawatt
NED	North–east–down
OCR	Operational capability review
ORE	Oblate rotating earth
ORR	Operational readiness review
PAC	Patriot advanced capability
PD	Pulse Doppler
PDR	Preliminary design review
PIP	Predicted-intercept-point

P_k	Probability of kill
PN	Proportional navigation
PRF	Pulse repetition frequency
PSD	Power spectral density
P_{ssk}	Probability of single-shot kill
RAM	Radar-absorbing material
RCS	radar cross section
RDR	Retirement and disposal review
RF	Radio frequency
RFP	Request for proposal
S	Shoot
SAR	Semi-active radar
SCU	Servo control unit
SDD	System design documents
SE	Systems engineering
SEMP	Systems engineering management plan
SER	Systems engineering review
SLS	Shoot–Look–Shoot
S/N	Signal-to-noise
SOS	System of systems
SR	Short range
SRBM	Short-range ballistic missile
SRD	Systems requirement document
SRR	Systems requirements review
SS	Shoot–Shoot
SSD	Ship self defense
SSS	Shoot–Shoot–Shoot
SST	Sensor suite tools
STC	Sensitivity time control
TDD	Target detecting device
TGO	Time to go
THL	Terminal homing time
THT	Terminal homing time
TLA	Top-level architecture
TLR	Top-level requirement
T/R	Transmit/receive
TRR	Test readiness review
TST	Target system tools
TVC	Thrust vector control
TVM	Track via missile
TWT	Traveling wave tube
VV&A	Validation, verification, and accreditation
W	Watt
WFG	Waveform generator

Index

A

Active radar (AR) seeker design
   acquisition performance, 119
   antenna gain, 115
   CPI, 116
   design considerations, 112
   design specifications, 118
   trade space, 113–114
Aeroprediction design, 131–132
Aim-point control, 52, 113
Air and missile defense (AMD) architecture, 53
Air and missile defense program plan
   background, 32–33
   definitions, 33–34
   measures of effectiveness
      contiguous coverage, 45–46
      continuous availability, 45
      DPTs, 40–44
      environmental resistance, 44–45
      firepower, 39–40
      reaction time, 35–39
   needs, 33–34
   preliminary design process, 32
   top-level requirements, 46
Air defense environment, 27–29
Airframe requirement design, 124–125
AMD down selection stoplight map, 196–197
Angle tracking, 110–112
Architecture functional requirements, 50
Area defense, 49
Attitude response requirements design, 138–140
   flight control design I, 141–149
   flight control design II, 149–154
   guidance and control requirements, 140
   guidance law design, 154–166
Augmented proportional navigation (APN), 164–165
Autopilot, 139, 144, 205
Azimuth raid, 26–27

B

Ballistic missile defense, 23
Battle management system (BMS), 52, 54
Battlespace analysis
   AMD down selection stoplight map, 196
   depth-of-fire performance
      interceptor flyout times, 173, 175
      summary plots, 179–185
   interceptor evaluation map, 196
   Monte Carlo engagement results, 192–193, 195
   notional engagement preliminary design study, 194
Battlespace Engineering Assessment Tool (BEAT), 93
Blind velocities, 235
Burnthrough techniques, 29

C

Central defense system (CDS), 50, 59–61
   handover, 65–68
   midcourse guidance, 61–65
   seeker pointing angle error, 68–70
Clutter
   air defense environment, 27–28
   definition, 225
   maximum range, 233
   processing considerations, 230
   radar element requirements, 202
   radar range equation, 225
   rejection degradation, 233–234
   surface clutter model, 225–227
   visibility, 235–236
   volume clutter, 228–230
Coherent processing interval (CPI), 230–232
Communication link system (CLS), 50, 52
Concept of operations (CONOPS), 12–13

248 Index

Configuration control management (CCM), 22
Configuration design, 125–130
Counterdetection and track phase design options, 42–43
Counterengagement and missile phase design options, 43–44
Counterpoint defense phase design options, 44
Countersurveillance and search phase design options, 41–42
CPI, *see* Coherent processing interval
Cruise missile defense, 23
Cuban Missile Crisis, 2

D

Deceptive jamming, 27
Defense penetration techniques (DPTs)
 counterdetection and track phase, 42–43
 counterengagement and missile phase, 43–44
 counterpoint defense phase, 44
 countersurveillance and search phase, 41–42
 definition, 40
 options, 41
Depth-of-fire performance, battlespace analysis
 interceptor flyout times, 173, 175
 summary plots, 179–185
Digital beamforming, 202
Directed energy weapon (DEW), 24–25
Dithering autopilot, 144

E

Eclipsing, 235
Engagement analysis design, 102–103, 105–106
Engagement computer system (ECS), 51, 54–55
Engagement system (ES)
 boundary requirements, 86
 Mach–altitude engagement envelope, 87–88
 specification development, 88–89

Environmental resistance, 44–45
Explicit guidance, 72

F

FCS, *see* Fire-control solution; Flight control system
Fill pulses, 232–233
Fire-control solution (FCS), 54, 56
Firepower, 39–40
Flight control design I
 instrumentation, 145–149
 servo actuator system, 143–145
 steering policy, 141–143
Flight control design II
 Bode plot, 154–155
 closed-loop step response, 154–155
 frequency domain requirements, 150
 root locus gain sensitivity, 153
 system architecture, 151–153
 time domain requirements, 149–150
Flight control system (FCS), 139, 151, 154, 161, 205
Frequency selective surface (FSS), 26

G

German V2 missile, 1
Glint, 238–239
Guidance integrated airframe (GIA), 161–162
Guidance integrated control (GIC), 161
Guidance integrated noise (GIN), 157–159
Guidance integrated target (GIT), 159–161
Guidance law design, 154–166

H

Handover error, 158–159
Heading error (HE), 65–66, 158, 204
High-energy weapons (HEWs), 51

I

Inertial reference unit (IRU) package, 205
Instantaneous bandwidth (IBW), 199–200

Index

Intelligence, surveillance, and reconnaissance (ISR) system, 50
Interceptor evaluation map, 196
Interceptor performance requirements
   launch and flyout phase, 204–206
   terminal homing and guidance, 203–204
Interceptors and flight analysis, *see* WGS-84 oblate, rotating earth model
IRU package, *see* Inertial reference unit package

J

Jamming, 120–121, 202–203

K

Kalman–Bucy filtering (KBF) approach, 162
Keep-out volume concept, 59

L

Land-based radars, 25
Land clutter, 28
Low-order equivalent model (LOEM), 160

M

Mach–altitude engagement envelope, 87–88
Main beam jamming, 29, 203
Maneuver design permissive bounds (MDPBs), 160
Man-in-the-loop (MITL) communication, 55
Mass and inertia design, 130–131
Mass ratio, 134–135
Material properties design, 138
Maximum clutter range, 233
Measures of effectiveness (MOEs), 14, 31
   contiguous coverage, 45–46
   continuous availability, 45
   DPTs, 40–44
   environmental resistance, 44–45
   firepower, 39–40
   reaction time, 35–39
Missile defense problem
   azimuth raid, 26–27
   DEWs, 24–25
   FSS, 26
   maximum unobstructed range, 25
   offensive missile speed, 24
   penetration techniques, 23
   radar detection range, 25
   stream raid, 26
Missile defense systems, need for, 1–2
Missile DATCOM, 132–133
Missile seekers, 26, 29, 105
   active radar (AR) design
      acquisition performance, 119
      antenna gain, 115
      CPI, 116
      design considerations, 112
      design specifications, 118
      trade space, 113–114
   angle tracking design, 110–112
   jamming design, 120–121
   signal transmission losses, 119–120
MOEs, *see* Measures of effectiveness
Moving target indicator (MTI), 28, 95, 199, 230, 233–235
Multipath, 27, 158, 202

N

Noise jamming, 27
Nonlinear guidance (NLG) law, 166
Non-Rayleigh target, 120
Nuclear-tipped Jupiter missiles, 1

O

Offensive missile speed, 24
Organic AMD systems, 52
ORE model, *see* WGS-84 oblate, rotating earth model

P

Patriot air defense missile system, 49
Phase A management plan, 8–9
Phase B management plan, 9–10

Point defense, 49
Predicted intercept point (PIP), 43, 51, 54–55, 65, 70–72
Preliminary design
  aeroprediction, 131–132
  airframe requirements, 124–125
  attitude response requirements, 138–140
    flight control design I, 141–149
    flight control design II, 149–154
    guidance and control requirements, 140
    guidance law design, 154–166
  configuration design, 125–130
  engagement analysis, 102–103, 105–106
  mass and inertia design, 130–131
  material properties design, 138
  missile seeker, 107–109
    active radar (AR) design, 112–119
    angle tracking, 110–112
    jamming, 120–121
    signal transmission losses, 119–120
  propulsion design, 133–137
  SST, 95–97
  translational and attitude response design, 121–124
  TST, 95
Pre-Phase A management plan, 7–8; *see also* Air and missile defense program plan
PRI, *see* Pulse repetition interval
Proportional navigation (PN), 163
Propulsion design
  burn time and vacuum thrust analysis, 135–136
  design parameters, 135
  mass ratio calculation, 134
  one-dimensional flyout design, 137
  SRM, 133
  thrust, 134
  vehicle mass analysis, 135–136
  velocity gain equation, 134
Pulse Doppler (PD), 28
Pulse repetition frequency (PRF), 235

Pulse repetition interval (PRI), 230
Pure proportional navigation (PPN), 163–164

Q

Quaternions, 224

R

Radar performance requirements
  antenna beamwidth, 200
  IBW, 199–200
  target signal-to-noise ratio, 202–203
Radar returns, 225
Radome boresight error (RBE), 158, 239–240
Radome boresight error slope (RBS), 239
Raids, 24
Range–velocity visibility, 234–235
Rayleigh target, 120
RBE, see Radome boresight error (RBE)
RBS, *see* Radome boresight error slope
Reaction time trade space, 60
Realistic true proportional navigation (R-TPN), 164
Receiver noise, 236–240
Requirement-driven acquisitions, 14–16
Risk management process, 22
Rocket equation, 134

S

Salvo time, 39
Seeker noise sources
  glint, 238–239
  RBS, 239–240
  receiver noise, 236–238
Semi-active radar (SAR) missile systems, 58
Sensor suite system (SSS), 56
Sensor suite tool (SST), 95–97
Shipborne radar, 25
Sidelobe jamming, 28–29
Signal transmission losses, 119–120
Solid rocket motors (SRMs), 133
SST, *see* Sensor suite tool
Stream raid, 26

Index

Surface clutter returns, 28, 225–227
System-of-systems (SOS) architecture, 14
System requirements
  CDS, 59–61
    handover, 65–68
    midcourse guidance, 61–65
    seeker pointing angle error, 68–70
  document, 90
  engagement system
    boundary requirements, 86
    Mach–altitude engagement envelope, 87–88
    specification development, 88–89
  keep-out volume concept, 59
  radar system, architecture of, 82–83
Systems engineering (SE)
  definitions for, 3
  phases, 6
  product realization responsibilities, 6
  systems architectural design responsibilities, 5–6
  technical evaluation responsibilities, 6
  technical management responsibilities, 5
  three dimensions of, 4
Systems engineering management plan (SEMP), 5, 17–19

T

Target and clutter returns
  blind velocities, 235
  clutter rejection degradation, 233–234
  CPI, 230–232
  eclipsing, 235
  fill pulses, 232–233
  maximum clutter range, 233
  radar returns, 225
  surface clutter returns, 225–227
  volume clutter returns, 228–230
Target system tool (TST), 95
Terrain bounce jamming (TBJ), 29
Theater defense, 49
Theoretical kill probability, 195
Three-axis body-fixed accelerometers, 147
Thrust vector control (TVC) actuator systems, 204–205
Time-dependent thrust, 133
Time-phased defense penetration design options, 41
Time-to-go (TGO) estimation errors, 43, 158–159
Towed decoy, 29
Translational and attitude response design, 121–124
True proportional navigation (TPN), 163–164
TST, *see* Target system tool

V

Verification and validation, 17
Volume clutter, 28, 225, 228
Volume clutter returns
  elliptically shaped antenna beam, 228–230
  non-beam-filled volume clutter, 229
  processing considerations, 230

W

Weapon system time constant, 43
WGS-84 oblate, rotating earth (ORE) model
  forces and moments, 220–224
  geocentric and geodetic latitude mathematical relationship, 216–217
    vehicle position, 217–218
  latitude, longitude, and altitude, 219–220
  oblateness effects, 213–216
  transformation matrices
    coordinate frames and position, 208–212
    velocity and acceleration, 212–213